they do this by thinking about business from first principles instead of formulas.

This book stems from a course about startups that I taught at Stanford in 2012. College students can become extremely skilled at a few specialties, but many never learn what to do with those skills in the wider world. My primary goal in teaching the class was to help my students see beyond the tracks laid down by academic specialties to the broader future that is theirs to create. One of those students, Blake Masters, took detailed class notes, which circulated far beyond the campus, and in *Zero to One* I have worked with him to revise the notes for a wider audience. There's no reason why the future should happen only at Stanford, or in college, or in Silicon Valley.

# ZERO

## TO

# ONE

# ZERO
## TO
## ONE

NOTES ON STARTUPS,

OR

HOW TO BUILD THE FUTURE

# PETER THIEL
## WITH BLAKE MASTERS

7 9 10 8 6

Virgin Books, an imprint of Ebury Publishing,
20 Vauxhall Bridge Road,
London SW1V 2SA

Virgin Books is part of the Penguin Random House group of companies
whose addresses can be found at global.penguinrandomhouse.com

Penguin
Random House
UK

First published in the United Kingdom by Virgin Books in 2014
First published in the United States by Crown Business in 2014
This edition first published in the United Kingdom by Virgin Books in 2015

www.eburypublishing.co.uk

A CIP catalogue record for this book is available from the British Library

ISBN: 9780753555200

Printed and bound by CPI Group (UK) Ltd, Croydon CR0 4YY

MIX
Paper from
responsible sources
FSC® C016897

Penguin Random House is committed to a sustainable future
for our business, our readers and our planet. This book is made
from Forest Stewardship Council® certified paper.

# Contents

# ZERO

### TO

# ONE

*Preface*

# ZERO TO ONE

EVERY MOMENT IN BUSINESS happens only once. The next Bill Gates will not build an operating system. The next Larry Page or Sergey Brin won't make a search engine. And the next Mark Zuckerberg won't create a social network. If you are copying these guys, you aren't learning from them.

Of course, it's easier to copy a model than to make something new. Doing what we already know how to do takes the world from 1 to *n,* adding more of something familiar. But every time we create something new, we go from 0 to 1. The act of creation is singular, as is the moment of creation, and the result is something fresh and strange.

Unless they invest in the difficult task of creating new things, companies will fail in the future no matter how big their profits remain today. What happens when we've gained everything to be had from fine-tuning the old lines of business that we've inherited? Unlikely as it sounds, the answer threatens to be far worse than the crisis of 2008. Today's "best practices" lead to dead ends; the best paths are new and untried.

In a world of gigantic administrative bureaucracies both public and private, searching for a new path might seem like hoping for a miracle. Actually, if American business is going to succeed, we are going to need hundreds, or even thousands, of miracles. This would be depressing but for one crucial fact: humans are distinguished from other species by our ability to work miracles. We call these miracles *technology*.

Technology is miraculous because it allows us to do *more with less*, ratcheting up our fundamental capabilities to a higher level. Other animals are instinctively driven to build things like dams or honeycombs, but we are the only ones that can invent new things and better ways of making them. Humans don't decide what to build by making choices from some cosmic catalog of options given in advance; instead, by creating new technologies, we rewrite the plan of the world. These are the kind of elementary truths we teach to second graders, but they are easy to forget in a world where so much of what we do is repeat what has been done before.

*Zero to One* is about how to build companies that create new things. It draws on everything I've learned directly as a co-founder of PayPal and Palantir and then an investor in hundreds of startups, including Facebook and SpaceX. But while I have noticed many patterns, and I relate them here, this book offers no formula for success. The paradox of teaching entrepreneurship is that such a formula necessarily cannot exist; because every innovation is new and unique, no authority can prescribe in concrete terms how to be innovative. Indeed, the single most powerful pattern I have noticed is that successful people find value in unexpected places, and

# 1

# THE CHALLENGE OF THE FUTURE

W HENEVER I INTERVIEW someone for a job, I like to ask this question: "What important truth do very few people agree with you on?"

This question sounds easy because it's straightforward. Actually, it's very hard to answer. It's intellectually difficult because the knowledge that everyone is taught in school is by definition agreed upon. And it's psychologically difficult because anyone trying to answer must say something she knows to be unpopular. Brilliant thinking is rare, but courage is in even shorter supply than genius.

Most commonly, I hear answers like the following:

"Our educational system is broken and urgently needs to be fixed."

"America is exceptional."

"There is no God."

Those are bad answers. The first and the second statements might be true, but many people already agree with them. The third statement simply takes one side in a familiar debate. A good answer takes the following form: "Most people believe in x, but the truth is the opposite of x." I'll give my own answer later in this chapter.

What does this contrarian question have to do with the future? In the most minimal sense, the future is simply the set of all moments yet to come. But what makes the future distinctive and important isn't that it hasn't happened yet, but rather that it will be a time when the world looks different from today. In this sense, if nothing about our society changes for the next 100 years, then the future is over 100 years away. If things change radically in the next decade, then the future is nearly at hand. No one can predict the future exactly, but we know two things: it's going to be different, and it must be rooted in today's world. Most answers to the contrarian question are different ways of seeing the present; good answers are as close as we can come to looking into the future.

## ZERO TO ONE:
## THE FUTURE OF PROGRESS

When we think about the future, we hope for a future of progress. That progress can take one of two forms. Horizontal or extensive progress means copying things that work—going from 1 to n. Horizontal progress is easy to imagine because we already know what it looks like. Vertical or intensive progress means doing new things—going from 0 to 1. Vertical progress is harder to imagine because it requires doing something nobody else has ever done. If you take one typewriter and build 100, you have made horizontal progress. If you have a typewriter and build a word processor, you have made vertical progress.

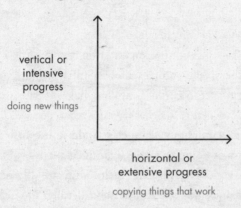

vertical or
intensive
progress

doing new things

horizontal or
extensive progress

copying things that work

At the macro level, the single word for horizontal progress is *globalization*—taking things that work somewhere and making them work everywhere. China is the paradigmatic example of globalization; its 20-year plan is to become like the United States is today. The Chinese have been straightforwardly copying everything that has worked in the developed world: 19th-century railroads, 20th-century air conditioning, and even entire cities. They might skip a few steps along the way—going straight to wireless without installing landlines, for instance—but they're copying all the same.

The single word for vertical, 0 to 1 progress is *technology*. The rapid progress of information technology in recent decades has made Silicon Valley the capital of "technology" in general. But there is no reason why technology should be

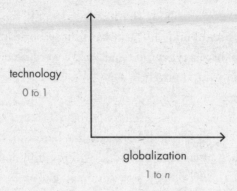

limited to computers. Properly understood, any new and better way of doing things is technology.

Because globalization and technology are different modes of progress, it's possible to have both, either, or neither at the same time. For example, 1815 to 1914 was a period of both rapid technological development and rapid globalization. Between the First World War and Kissinger's trip to reopen relations with China in 1971, there was rapid technological development but not much globalization. Since 1971, we have seen rapid globalization along with limited technological development, mostly confined to IT.

This age of globalization has made it easy to imagine that the decades ahead will bring more convergence and more sameness. Even our everyday language suggests we believe in a kind of technological end of history: the division of the world into the so-called developed and developing nations implies that the "developed" world has already achieved the achievable, and that poorer nations just need to catch up.

But I don't think that's true. My own answer to the contrarian question is that most people think the future of the world will be defined by globalization, but the truth is that technology matters more. Without technological change, if China doubles its energy production over the next two decades, it will also double its air pollution. If every one of India's hundreds of millions of households were to live the way Americans already do—using only today's tools—the result would be environmentally catastrophic. Spreading old ways to create wealth around the world will result in devastation, not riches. In a world of scarce resources, globalization without new technology is unsustainable.

New technology has never been an automatic feature of history. Our ancestors lived in static, zero-sum societies where success meant seizing things from others. They created new sources of wealth only rarely, and in the long run they could never create enough to save the average person from an extremely hard life. Then, after 10,000 years of fitful advance from primitive agriculture to medieval windmills and 16th-century astrolabes, the modern world suddenly experienced relentless technological progress from the advent of the steam engine in the 1760s all the way up to about 1970. As a result, we have inherited a richer society than any previous generation would have been able to imagine.

Any generation excepting our parents' and grandparents', that is: in the late 1960s, they expected this progress to continue. They looked forward to a four-day workweek, energy too cheap to meter, and vacations on the moon. But it didn't happen. The smartphones that distract us from our surroundings also distract us from the fact that our surroundings are strangely old: only computers and communications have improved dramatically since midcentury. That doesn't

mean our parents were wrong to imagine a better future—
they were only wrong to expect it as something automatic.
Today our challenge is to both imagine and create the new
technologies that can make the 21st century more peaceful
and prosperous than the 20th.

## STARTUP THINKING

New technology tends to come from new ventures—
startups. From the Founding Fathers in politics to the Royal
Society in science to Fairchild Semiconductor's "traitorous
eight" in business, small groups of people bound together
by a sense of mission have changed the world for the better.
The easiest explanation for this is negative: it's hard to develop new things in big organizations, and it's even harder
to do it by yourself. Bureaucratic hierarchies move slowly,
and entrenched interests shy away from risk. In the most dysfunctional organizations, signaling that work is being done
becomes a better strategy for career advancement than actually doing work (if this describes your company, you should
quit now). At the other extreme, a lone genius might create
a classic work of art or literature, but he could never create
an entire industry. Startups operate on the principle that you
need to work with other people to get stuff done, but you
also need to stay small enough so that you actually can.

Positively defined, a startup is the largest group of people
you can convince of a plan to build a different future. A new
company's most important strength is new thinking: even
more important than nimbleness, small size affords space to

think. This book is about the questions you must ask and answer to succeed in the business of doing new things: what follows is not a manual or a record of knowledge but an exercise in thinking. Because that is what a startup has to do: question received ideas and rethink business from scratch.

# 2

# PARTY LIKE IT'S 1999

OUR CONTRARIAN QUESTION—*What important truth do very few people agree with you on?*—is difficult to answer directly. It may be easier to start with a preliminary: what does everybody agree on? "Madness is rare in individuals—but in groups, parties, nations, and ages it is the rule," Nietzsche wrote (before he went mad). If you can identify a delusional popular belief, you can find what lies hidden behind it: the contrarian truth.

Consider an elementary proposition: companies exist to make money, not to lose it. This should be obvious to any thinking person. But it wasn't so obvious to many in the late 1990s, when no loss was too big to be described as an investment in an even bigger, brighter future. The conventional wisdom of the "New Economy" accepted page views as a more authoritative, forward-looking financial metric than something as pedestrian as profit.

Conventional beliefs only ever come to appear arbitrary and wrong in retrospect; whenever one collapses, we call the old belief a *bubble*. But the distortions caused by bubbles don't disappear when they pop. The internet craze of the '90s was the biggest bubble since the crash of 1929, and the lessons learned afterward define and distort almost all thinking about technology today. The first step to thinking clearly is to question what we think we know about the past.

## A QUICK HISTORY OF THE '90S

The 1990s have a good image. We tend to remember them as a prosperous, optimistic decade that happened to end with the internet boom and bust. But many of those years were not as cheerful as our nostalgia holds. We've long since forgotten the global context for the 18 months of dot-com mania at decade's end.

The '90s started with a burst of euphoria when the Berlin Wall came down in November '89. It was short-lived. By mid-1990, the United States was in recession. Technically the downturn ended in March '91, but recovery was slow and unemployment continued to rise until July '92. Manufacturing never fully rebounded. The shift to a service economy was protracted and painful.

1992 through the end of 1994 was a time of general malaise. Images of dead American soldiers in Mogadishu looped on cable news. Anxiety about globalization and U.S. competitiveness intensified as jobs flowed to Mexico. This pessimistic undercurrent drove then-president Bush 41 out of office and won Ross Perot nearly 20% of the popular vote in '92—the best showing for a third-party candidate since

Theodore Roosevelt in 1912. And whatever the cultural fascination with Nirvana, grunge, and heroin reflected, it wasn't hope or confidence.

Silicon Valley felt sluggish, too. Japan seemed to be winning the semiconductor war. The internet had yet to take off, partly because its commercial use was restricted until late 1992 and partly due to the lack of user-friendly web browsers. It's telling that when I arrived at Stanford in 1985, economics, not computer science, was the most popular major. To most people on campus, the tech sector seemed idiosyncratic or even provincial.

The internet changed all this. The Mosaic browser was officially released in November 1993, giving regular people a way to get online. Mosaic became Netscape, which released its Navigator browser in late 1994. Navigator's adoption grew so quickly—from about 20% of the browser market in January 1995 to almost 80% less than 12 months later—that Netscape was able to IPO in August '95 even though it wasn't yet profitable. Within five months, Netscape stock had shot up from $28 to $174 per share. Other tech companies were booming, too. Yahoo! went public in April '96 with an $848 million valuation. Amazon followed suit in May '97 at $438 million. By spring of '98, each company's stock had more than quadrupled. Skeptics questioned earnings and revenue multiples higher than those for any non-internet company. It was easy to conclude that the market had gone crazy.

This conclusion was understandable but misplaced. In December '96—more than three years before the bubble actually burst—Fed chairman Alan Greenspan warned that "irrational exuberance" might have "unduly escalated asset values." Tech investors were exuberant, but it's not clear that they were so irrational. It is too easy to forget that things weren't going very well in the rest of the world at the time.

The East Asian financial crises hit in July 1997. Crony capitalism and massive foreign debt brought the Thai, Indonesian, and South Korean economies to their knees. The ruble crisis followed in August '98 when Russia, hamstrung by chronic fiscal deficits, devalued its currency and defaulted on its debt. American investors grew nervous about a nation with 10,000 nukes and no money; the Dow Jones Industrial Average plunged more than 10% in a matter of days.

People were right to worry. The ruble crisis set off a chain reaction that brought down Long-Term Capital Management, a highly leveraged U.S. hedge fund. LTCM managed to lose $4.6 billion in the latter half of 1998, and still had over $100 billion in liabilities when the Fed intervened with a massive bailout and slashed interest rates in order to prevent systemic disaster. Europe wasn't doing that much better. The euro launched in January 1999 to great skepticism and apathy. It rose to $1.19 on its first day of trading but sank to $0.83 within two years. In mid-2000, G7 central bankers had to prop it up with a multibillion-dollar intervention.

So the backdrop for the short-lived dot-com mania that started in September 1998 was a world in which nothing else seemed to be working. The Old Economy couldn't handle the challenges of globalization. Something needed to work—and work in a big way—if the future was going to be better at all. By indirect proof, the New Economy of the internet was the only way forward.

## MANIA: SEPTEMBER 1998–MARCH 2000

Dot-com mania was intense but short—18 months of insanity from September 1998 to March 2000. It was a Silicon Valley gold rush: there was money everywhere, and no shortage of exuberant, often sketchy people to chase it. Every week, dozens of new startups competed to throw the most lavish launch party. (Landing parties were much more rare.) Paper millionaires would rack up thousand-dollar dinner bills and try to pay with shares of their startup's stock—sometimes it even worked. Legions of people decamped from their well-paying jobs to found or join startups. One 40-something grad student that I knew was running six different companies in 1999. (Usually, it's considered weird to be a 40-year-old graduate student. Usually, it's considered insane to start a half-dozen companies at once. But in the late '90s, people

## DOT-COM BOOM

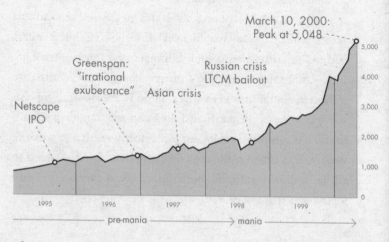

March 10, 2000:
Peak at 5,048

Greenspan:
"irrational
exuberance"

Russian crisis
LTCM bailout

Asian crisis

Netscape
IPO

1995     1996     1997     1998     1999

pre-mania ———→ mania ———→

could believe that was a winning combination.) Everybody should have known that the mania was unsustainable; the most "successful" companies seemed to embrace a sort of anti-business model where they *lost* money as they grew. But it's hard to blame people for dancing when the music was playing; irrationality was rational given that appending ".com" to your name could double your value overnight.

## PAYPAL MANIA

When I was running PayPal in late 1999, I was scared out of my wits—not because I didn't believe in our company, but because it seemed like everyone else in the Valley was ready to believe anything at all. Everywhere I looked, people were starting and flipping companies with alarming casualness. One acquaintance told me how he had planned an IPO from his living room before he'd even incorporated his company—and he didn't think that was weird. In this kind of environment, acting sanely began to seem eccentric.

At least PayPal had a suitably grand mission—the kind that post-bubble skeptics would later describe as grandiose: we wanted to create a new internet currency to replace the U.S. dollar. Our first product let people beam money from one PalmPilot to another. However, nobody had any use for that product except the journalists who voted it one of the 10 worst business ideas of 1999. PalmPilots were still too exotic then, but email was already commonplace, so we decided to create a way to send and receive payments over email.

By the fall of '99, our email payment product worked well—anyone could log in to our website and easily transfer money. But we didn't have enough customers, growth was slow, and expenses mounted. For PayPal to work, we needed to attract a critical mass of at least a million users. Advertising was too ineffective to justify the cost. Prospective deals with big banks kept falling through. So we decided to pay people to sign up.

We gave new customers $10 for joining, and we gave them $10 more every time they referred a friend. This got us hundreds of thousands of new customers and an exponential growth rate. Of course, this customer acquisition strategy was unsustainable on its own—when you pay people to be your customers, exponential growth means an exponentially growing cost structure. Crazy costs were typical at that time in the Valley. But we thought our huge costs were sane: given a large user base, PayPal had a clear path to profitability by taking a small fee on customers' transactions.

We knew we'd need more funding to reach that goal. We also knew that the boom was going to end. Since we didn't expect investors' faith in our mission to survive the coming crash, we moved fast to raise funds while we could. On February 16, 2000, the *Wall Street Journal* ran a story lauding our viral growth and suggesting that PayPal was worth $500 million. When we raised $100 million the next month, our lead investor took the *Journal*'s back-of-the-envelope valuation as authoritative. (Other investors were in even more of a hurry. A South Korean firm wired us $5 million without first negotiating a deal or signing any documents. When I tried to return the money, they wouldn't tell me where to send it.) That March 2000 financing round bought us the time we needed to make PayPal a success. Just as we closed the deal, the bubble popped.

## LESSONS LEARNED

*'Cause they say 2,000 zero zero party over, oops! Out of time!*
*So tonight I'm gonna party like it's 1999!*

—PRINCE

The NASDAQ reached 5,048 at its peak in the middle of March 2000 and then crashed to 3,321 in the middle of April. By the time it bottomed out at 1,114 in October 2002, the country had long since interpreted the market's collapse as a kind of divine judgment against the technological optimism of the '90s. The era of cornucopian hope was relabeled as an era of crazed greed and declared to be definitely over.

Everyone learned to treat the future as fundamentally

## DOT-COM BUST

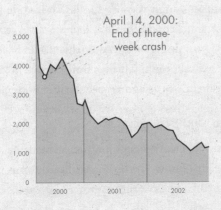

indefinite, and to dismiss as an extremist anyone with plans big enough to be measured in years instead of quarters. Globalization replaced technology as the hope for the future. Since the '90s migration "from bricks to clicks" didn't work as hoped, investors went back to bricks (housing) and BRICs (globalization). The result was another bubble, this time in real estate.

The entrepreneurs who stuck with Silicon Valley learned four big lessons from the dot-com crash that still guide business thinking today:

1. Make incremental advances

   Grand visions inflated the bubble, so they should not
   be indulged. Anyone who claims to be able to do
   something great is suspect, and anyone who wants
   to change the world should be more humble. Small,
   incremental steps are the only safe path forward.

2. Stay lean and flexible

   All companies must be "lean," which is code for
   "unplanned." You should not know what your
   business will do; planning is arrogant and inflexible.
   Instead you should try things out, "iterate," and treat
   entrepreneurship as agnostic experimentation.

3. Improve on the competition

   Don't try to create a new market prematurely. The
   only way to know you have a real business is to start
   with an already existing customer, so you should build

your company by improving on recognizable products already offered by successful competitors.

4. Focus on product, not sales

If your product requires advertising or salespeople to sell it, it's not good enough: technology is primarily about product development, not distribution. Bubble-era advertising was obviously wasteful, so the only sustainable growth is viral growth.

These lessons have become dogma in the startup world; those who would ignore them are presumed to invite the justified doom visited upon technology in the great crash of 2000. And yet the opposite principles are probably more correct:

1. *It is better to risk boldness than triviality.*

2. *A bad plan is better than no plan.*

3. *Competitive markets destroy profits.*

4. *Sales matters just as much as product.*

It's true that there was a bubble in technology. The late '90s was a time of hubris: people believed in going from 0 to 1. Too few startups were actually getting there, and many never went beyond talking about it. But people understood that we had no choice but to find ways to do more with less. The market high of March 2000 was obviously a peak of insanity; less obvious but more important, it was also a peak of clarity. People looked far into the future, saw how much valuable new technology we would need to get there safely, and judged themselves capable of creating it.

We still need new technology, and we may even need some 1999-style hubris and exuberance to get it. To build the next generation of companies, we must abandon the dogmas created after the crash. That doesn't mean the opposite ideas are automatically true: you can't escape the madness of crowds by dogmatically rejecting them. Instead ask yourself: how much of what you know about business is shaped by mistaken reactions to past mistakes? The most contrarian thing of all is not to oppose the crowd but to think for yourself.

# 3

## ALL HAPPY COMPANIES ARE DIFFERENT

THE BUSINESS VERSION of our contrarian question is: *what valuable company is nobody building*? This question is harder than it looks, because your company could create a lot of value without becoming very valuable itself. Creating value is not enough—you also need to capture some of the value you create.

This means that even very big businesses can be bad businesses. For example, U.S. airline companies serve millions of passengers and create hundreds of billions of dollars of value each year. But in 2012, when the average airfare each way was $178, the airlines made only 37 cents per passenger trip. Compare them to Google, which creates less value but captures far more. Google brought in $50 billion in 2012 (versus $160 billion for the airlines), but it kept 21% of those revenues as profits—more than 100 times the airline industry's profit margin that year. Google makes so much money that it's now worth three times more than every U.S. airline combined.

The airlines compete with each other, but Google stands alone. Economists use two simplified models to explain the difference: perfect competition and monopoly.

"Perfect competition" is considered both the ideal and the default state in Economics 101. So-called perfectly competitive markets achieve equilibrium when producer supply meets consumer demand. Every firm in a competitive market is undifferentiated and sells the same homogeneous products. Since no firm has any market power, they must all sell at whatever price the market determines. If there is money to be made, new firms will enter the market, increase supply, drive prices down, and thereby eliminate the profits that attracted them in the first place. If too many firms enter the market, they'll suffer losses, some will fold, and prices will rise back to sustainable levels. Under perfect competition, in the long run *no company makes an economic profit*.

The opposite of perfect competition is monopoly. Whereas a competitive firm must sell at the market price, a monopoly owns its market, so it can set its own prices. Since it has no competition, it produces at the quantity and price combination that maximizes its profits.

To an economist, every monopoly looks the same, whether it deviously eliminates rivals, secures a license from the state, or innovates its way to the top. In this book, we're not interested in illegal bullies or government favorites: by "monopoly," we mean the kind of company that's so good at what it does that no other firm can offer a close substitute. Google is a good example of a company that went from 0 to 1: it hasn't competed in search since the early 2000s, when it definitively distanced itself from Microsoft and Yahoo!

Americans mythologize competition and credit it with saving us from socialist bread lines. Actually, capitalism and competition are opposites. Capitalism is premised on the accumulation of capital, but under perfect competition all profits get competed away. The lesson for entrepreneurs is clear: *if you want to create and capture lasting value, don't build an undifferentiated commodity business.*

## LIES PEOPLE TELL

How much of the world is actually monopolistic? How much is truly competitive? It's hard to say, because our common conversation about these matters is so confused. To the outside observer, all businesses can seem reasonably alike, so it's easy to perceive only small differences between them.

### PERCEPTION:
### FIRMS ARE SIMILAR

But the reality is much more binary than that. There's an enormous difference between perfect competition and monopoly, and most businesses are much closer to one extreme than we commonly realize.

# REALITY:
# DIFFERENCES ARE DEEP

The confusion comes from a universal bias for describing market conditions in self-serving ways: both monopolists and competitors are incentivized to bend the truth.

## *Monopoly Lies*

Monopolists lie to protect themselves. They know that bragging about their great monopoly invites being audited, scrutinized, and attacked. Since they very much want their monopoly profits to continue unmolested, they tend to do whatever they can to conceal their monopoly—usually by exaggerating the power of their (nonexistent) competition.

Think about how Google talks about its business. It certainly doesn't *claim* to be a monopoly. But is it one? Well, it depends: a monopoly in *what*? Let's say that Google is primarily a search engine. As of May 2014, it owns about 68% of the search market. (Its closest competitors, Microsoft and Yahoo!, have about 19% and 10%, respectively.) If that doesn't seem dominant enough, consider the fact that the word "google" is now an official entry in the *Oxford English Dictionary*—as a verb. Don't hold your breath waiting for that to happen to Bing.

But suppose we say that Google is primarily an advertising company. That changes things. The U.S. search engine ad-

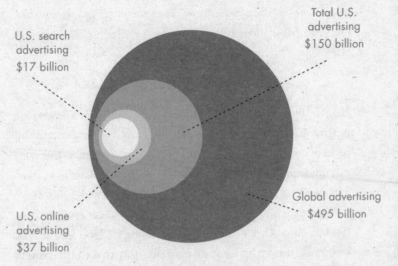

U.S. search
advertising
$17 billion

Total U.S.
advertising
$150 billion

U.S. online
advertising
$37 billion

Global advertising
$495 billion

vertising market is $17 billion annually. Online advertising is $37 billion annually. The entire U.S. advertising market is $150 billion. And *global* advertising is a $495 billion market. So even if Google completely monopolized U.S. search engine advertising, it would own just 3.4% of the global advertising market. From this angle, Google looks like a small player in a competitive world.

What if we frame Google as a multifaceted technology company instead? This seems reasonable enough; in addition to its search engine, Google makes dozens of other software products, not to mention robotic cars, Android phones, and wearable computers. But 95% of Google's revenue comes from search advertising; its other products generated just $2.35 billion in 2012, and its consumer tech products a mere fraction of that. Since consumer tech is a $964 billion market globally, Google owns less than 0.24% of it—a far cry from relevance, let alone monopoly. Framing itself as just another tech company allows Google to escape all sorts of unwanted attention.

## Competitive Lies

Non-monopolists tell the opposite lie: "we're in a league of our own." Entrepreneurs are always biased to understate the scale of competition, but that is the biggest mistake a startup can make. The fatal temptation is to describe your market extremely narrowly so that you dominate it by definition.

Suppose you want to start a restaurant that serves British food in Palo Alto. "No one else is doing it," you might reason. "We'll own the entire market." But that's only true if the relevant market is the market for British food specifically. What if the actual market is the Palo Alto restaurant market in general? And what if all the restaurants in nearby towns are part of the relevant market as well?

These are hard questions, but the bigger problem is that you have an incentive not to ask them at all. When you hear that most new restaurants fail within one or two years, your instinct will be to come up with a story about how yours is different. You'll spend time trying to convince people that you are exceptional instead of seriously considering whether that's true. It would be better to pause and consider whether there are people in Palo Alto who would rather eat British food above all else. It's very possible they don't exist.

In 2001, my co-workers at PayPal and I would often get lunch on Castro Street in Mountain View. We had our pick of restaurants, starting with obvious categories like Indian, sushi, and burgers. There were more options once we settled on a type: North Indian or South Indian, cheaper or fancier, and so on. In contrast to the competitive local restaurant market, PayPal was at that time the only email-based payments company in the world. We employed fewer people than the restaurants on Castro Street did, but our business was much more valuable than all of those restaurants combined. Starting a new South Indian restaurant is a really hard way to make money. If you lose sight of competitive reality and focus on trivial differentiating factors—maybe you think your naan is superior because of your great-grandmother's recipe—your business is unlikely to survive.

Creative industries work this way, too. No screenwriter wants to admit that her new movie script simply rehashes what has already been done before. Rather, the pitch is: "This film will combine various exciting elements in entirely new ways." It could even be true. Suppose her idea is to have

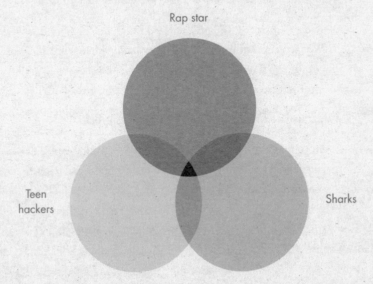

Jay-Z star in a cross between *Hackers* and *Jaws:* rap star joins elite group of hackers to catch the shark that killed his friend. *That* has definitely never been done before. But, like the lack of British restaurants in Palo Alto, maybe that's a good thing.

Non-monopolists exaggerate their distinction by defining their market as the *intersection* of various smaller markets:

> British food ∩ restaurant ∩ Palo Alto
>
> Rap star ∩ hackers ∩ sharks

Monopolists, by contrast, disguise their monopoly by framing their market as the *union* of several large markets:

> search engine ∪ mobile phones ∪ wearable
> computers ∪ self-driving cars

What does a monopolist's union story look like in practice? Consider a statement from Google chairman Eric Schmidt's testimony at a 2011 congressional hearing:

> We face an extremely competitive landscape in which consumers have a multitude of options to access information.

Or, translated from PR-speak to plain English:

> Google is a small fish in a big pond. We could be swallowed whole at any time. We are not the monopoly that the government is looking for.

## RUTHLESS PEOPLE

The problem with a competitive business goes beyond lack of profits. Imagine you're running one of those restaurants in Mountain View. You're not that different from dozens of your competitors, so you've got to fight hard to survive. If you offer affordable food with low margins, you can probably pay employees only minimum wage. And you'll need to squeeze out every efficiency: that's why small restaurants put Grandma to work at the register and make the kids wash dishes in the back. Restaurants aren't much better even at the very highest rungs, where reviews and ratings like Michelin's star system enforce a culture of intense competition that can drive chefs crazy. (French chef and winner of three Michelin stars Bernard Loiseau was quoted as saying, "If I lose a star, I will commit suicide." Michelin maintained his rating, but Loiseau killed himself anyway in 2003 when a competing French dining guide downgraded his restaurant.) The competitive ecosystem pushes people toward ruthlessness or death.

A monopoly like Google is different. Since it doesn't have to worry about competing with anyone, it has wider latitude to care about its workers, its products, and its impact on the wider world. Google's motto—"Don't be evil"—is in part a branding ploy, but it's also characteristic of a kind of business that's successful enough to take ethics seriously without jeopardizing its own existence. In business, *money is either an important thing or it is everything*. Monopolists can afford to think about things other than making money; non-monopolists can't. In perfect competition, a business is so focused on to-day's margins that it can't possibly plan for a long-term future. Only one thing can allow a business to transcend the daily brute struggle for survival: monopoly profits.

## MONOPOLY CAPITALISM

So, a monopoly is good for everyone on the inside, but what about everyone on the outside? Do outsized profits come at the expense of the rest of society? Actually, yes: profits come out of customers' wallets, and monopolies deserve their bad reputation—*but only in a world where nothing changes.*

In a static world, a monopolist is just a rent collector. If you corner the market for something, you can jack up the price; others will have no choice but to buy from you. Think of the famous board game: deeds are shuffled around from player to player, but the board never changes. There's no way to win by inventing a better kind of real estate development. The relative values of the properties are fixed for all time, so all you can do is try to buy them up.

But the world we live in is dynamic: it's possible to invent new and better things. Creative monopolists give customers *more* choices by adding entirely new categories of abundance to the world. Creative monopolies aren't just good for the rest of society; they're powerful engines for making it better.

Even the government knows this: that's why one of its departments works hard to create monopolies (by granting patents to new inventions) even though another part hunts them down (by prosecuting antitrust cases). It's possible to question whether anyone should really be awarded a *legally enforceable* monopoly simply for having been the first to think of something like a mobile software design. But it's clear that something like Apple's monopoly profits from designing, producing, and marketing the iPhone were the reward for creating greater abundance, not artificial scarcity: customers were happy to finally have the choice of paying high prices to get a smartphone that actually works.

The dynamism of new monopolies itself explains why old monopolies don't strangle innovation. With Apple's iOS at the forefront, the rise of mobile computing has dramatically reduced Microsoft's decades-long operating system dominance. Before that, IBM's hardware monopoly of the '60s and '70s was overtaken by Microsoft's software monopoly. AT&T had a monopoly on telephone service for most of the 20th century, but now anyone can get a cheap cell phone plan from any number of providers. If the tendency of monopoly businesses were to hold back progress, they would be dangerous and we'd be right to oppose them. But the history of progress is a history of better monopoly businesses replacing incumbents.

Monopolies drive progress because the promise of years or even decades of monopoly profits provides a powerful incentive to innovate. Then monopolies can keep innovating because profits enable them to make the long-term plans and to finance the ambitious research projects that firms locked in competition can't dream of.

So why are economists obsessed with competition as an ideal state? It's a relic of history. Economists copied their mathematics from the work of 19th-century physicists: they see individuals and businesses as interchangeable atoms, not as unique creators. Their theories describe an equilibrium state of perfect competition because that's what's easy to model, not because it represents the best of business. But it's worth recalling that the long-run equilibrium predicted by 19th-century physics was a state in which all energy is evenly distributed and everything comes to rest—also known as the heat death of the universe. Whatever your views on thermodynamics, it's a powerful metaphor: in business, equilibrium means stasis, and stasis means death. If your industry is in a competitive equilibrium, the death of your business won't matter to the world; some other undifferentiated competitor will always be ready to take your place.

Perfect equilibrium may describe the void that is most of the universe. It may even characterize many businesses. But every new creation takes place far from equilibrium. In the real world outside economic theory, every business is successful exactly to the extent that it does something others cannot. Monopoly is therefore not a pathology or an exception. *Monopoly is the condition of every successful business.*

Tolstoy opens *Anna Karenina* by observing: "All happy families are alike; each unhappy family is unhappy in its own way." Business is the opposite. All happy companies are different: each one earns a monopoly by solving a unique problem. All failed companies are the same: they failed to escape competition.

# 4

## THE IDEOLOGY OF COMPETITION

CREATIVE MONOPOLY means new products that benefit everybody and sustainable profits for the creator. Competition means no profits for anybody, no meaningful differentiation, and a struggle for survival. So why do people believe that competition is healthy? The answer is that competition is not just an economic concept or a simple inconvenience that individuals and companies must deal with in the marketplace. More than anything else, competition is an ideology—*the* ideology—that pervades our society and distorts our thinking. We preach competition, internalize its necessity, and enact its commandments; and as a result, we trap ourselves within it—even though the more we compete, the less we gain.

This is a simple truth, but we've all been trained to ignore it. Our educational system both drives and reflects our obsession with competition. Grades themselves allow precise measurement of each student's competitiveness; pupils with the highest marks receive status and credentials. We teach every young person the same subjects in mostly the same ways, irrespective of individual talents and preferences. Students who don't learn best by sitting still at a desk are made to feel somehow inferior, while children who excel on conventional measures like tests and assignments end up defining their identities in terms of this weirdly contrived academic parallel reality.

And it gets worse as students ascend to higher levels of the tournament. Elite students climb confidently until they reach a level of competition sufficiently intense to beat their dreams out of them. Higher education is the place where people who had big plans in high school get stuck in fierce rivalries with equally smart peers over conventional careers like management consulting and investment banking. For the privilege of being turned into conformists, students (or their families) pay hundreds of thousands of dollars in skyrocketing tuition that continues to outpace inflation. Why are we doing this to ourselves?

I wish I had asked myself when I was younger. My path was so tracked that in my 8th-grade yearbook, one of my friends predicted—accurately—that four years later I would enter Stanford as a sophomore. And after a conventionally successful undergraduate career, I enrolled at Stanford Law School, where I competed even harder for the standard badges of success.

The highest prize in a law student's world is unambiguous: out of tens of thousands of graduates each year, only a few dozen get a Supreme Court clerkship. After clerking on a federal appeals court for a year, I was invited to interview for clerkships with Justices Kennedy and Scalia. My meetings with the Justices went well. I was so close to winning this last competition. If only I got the clerkship, I thought, I would be set for life. But I didn't. At the time, I was devastated.

In 2004, after I had built and sold PayPal, I ran into an old friend from law school who had helped me prepare my failed clerkship applications. We hadn't spoken in nearly a decade. His first question wasn't "How are you doing?" or "Can you believe it's been so long?" Instead, he grinned and asked: "So, Peter, aren't you glad you didn't get that clerkship?" With the benefit of hindsight, we both knew that winning that ultimate competition would have changed my life for the worse. Had I actually clerked on the Supreme Court, I probably would have spent my entire career taking depositions or drafting other people's business deals instead of creating anything new. It's hard to say how much would be different, but the opportunity costs were enormous. All Rhodes Scholars had a great future in their past.

## WAR AND PEACE

Professors downplay the cutthroat culture of academia, but managers never tire of comparing business to war. MBA students carry around copies of Clausewitz and Sun Tzu. War metaphors invade our everyday business language: we use *headhunters* to build up a sales *force* that will enable us to take a *captive market* and *make a killing*. But really it's competition, not business, that is like war: allegedly necessary, supposedly valiant, but ultimately destructive.

Why do people compete with each other? Marx and Shakespeare provide two models for understanding almost every kind of conflict.

37

According to Marx, people fight because they are different. The proletariat fights the bourgeoisie because they have completely different ideas and goals (generated, for Marx, by their very different material circumstances). The greater the differences, the greater the conflict.

To Shakespeare, by contrast, all combatants look more or less alike. It's not at all clear why they should be fighting, since they have nothing to fight about. Consider the opening line from *Romeo and Juliet:* "Two households, both alike in dignity." The two houses are alike, yet they hate each other. They grow even more similar as the feud escalates. Eventually, they lose sight of why they started fighting in the first place.

In the world of business, at least, Shakespeare proves the superior guide. Inside a firm, people become obsessed with their competitors for career advancement. Then the firms themselves become obsessed with their competitors in the marketplace. Amid all the human drama, people lose sight of what matters and focus on their rivals instead.

Let's test the Shakespearean model in the real world. Imagine a production called *Gates and Schmidt,* based on *Romeo and Juliet.* Montague is Microsoft. Capulet is Google. Two great families, run by alpha nerds, sure to clash on account of their sameness.

As with all good tragedy, the conflict seems inevitable only in retrospect. In fact it was entirely avoidable. These families came from very different places. The House of Montague built operating systems and office applications. The House of Capulet wrote a search engine. What was there to fight about?

Lots, apparently. As a startup, each clan had been content to leave the other alone and prosper independently. But as they grew, they began to focus on each other. Montagues obsessed about Capulets obsessed about Montagues. The result? Windows vs. Chrome OS, Bing vs. Google Search, Explorer vs. Chrome, Office vs. Docs, and Surface vs. Nexus.

Just as war cost the Montagues and Capulets their children, it cost Microsoft and Google their dominance: Apple came along and overtook them all. In January 2013, Apple's market capitalization was $500 billion, while Google and Microsoft combined were worth $467 billion. Just three years before, Microsoft and Google were *each* more valuable than Apple. War is costly business.

Rivalry causes us to overemphasize old opportunities and slavishly copy what has worked in the past. Consider the recent proliferation of mobile credit card readers. In October 2010, a startup called Square released a small, white, square-shaped product that let anyone with an iPhone swipe and accept credit cards. It was the first good payment processing solution for mobile handsets. Imitators promptly sprang into

action. A Canadian company called NetSecure launched its own card reader in a half-moon shape. Intuit brought a cylindrical reader to the geometric battle. In March 2012, eBay's PayPal unit launched its own copycat card reader. It was shaped like a triangle—a clear jab at Square, as three sides are simpler than four. One gets the sense that this Shakespearean saga won't end until the apes run out of shapes.

The hazards of imitative competition may partially explain why individuals with an Asperger's-like social ineptitude seem to be at an advantage in Silicon Valley today. If you're less sensitive to social cues, you're less likely to do the same things as everyone else around you. If you're interested in making things or programming computers, you'll be less afraid to pursue those activities single-mindedly and thereby become incredibly good at them. Then when you apply your skills, you're a little less likely than others to give up your own convictions: this can save you from getting caught up in crowds competing for obvious prizes.

Competition can make people hallucinate opportunities where none exist. The crazy '90s version of this was the fierce battle for the online pet store market. It was Pets.com vs. PetStore.com vs. Petopia.com vs. what seemed like dozens of others. Each company was obsessed with defeating its rivals, precisely because there were no substantive differences to focus on. Amid all the tactical questions—Who could price chewy dog toys most aggressively? Who could create the best Super Bowl ads?—these companies totally lost sight of the wider question of whether the online pet supply market was the right space to be in. Winning is better than losing, but everybody loses when the war isn't one worth fighting.

When Pets.com folded after the dot-com crash, $300 million of investment capital disappeared with it.

Other times, rivalry is just weird and distracting. Consider the Shakespearean conflict between Larry Ellison, cofounder and CEO of Oracle, and Tom Siebel, a top salesman at Oracle and Ellison's protégé before he went on to found Siebel Systems in 1993. Ellison was livid at what he thought was Siebel's betrayal. Siebel hated being in the shadow of his former boss. The two men were basically identical—hard-charging Chicagoans who loved to sell and hated to lose—so their hatred ran deep. Ellison and Siebel spent the second half of the '90s trying to sabotage each other. At one point, Ellison sent truckloads of ice cream sandwiches to Siebel's headquarters to try to convince Siebel employees to jump ship. The copy on the wrappers? "Summer is near. Oracle is here. To brighten your day and your career."

Strangely, Oracle intentionally accumulated enemies. Ellison's theory was that it's always good to have an enemy, so long as it was large enough to *appear* threatening (and thus motivational to employees) but not so large as to actually threaten the company. So Ellison was probably thrilled when in 1996 a small database company called Informix put up a billboard near Oracle's Redwood Shores headquarters that read: CAUTION: DINOSAUR CROSSING. Another Informix billboard on northbound Highway 101 read: YOU'VE JUST PASSED REDWOOD SHORES. SO DID WE.

Oracle shot back with a billboard that implied that Informix's software was slower than snails. Then Informix CEO Phil White decided to make things personal. When White learned that Larry Ellison enjoyed Japanese samurai culture, he commissioned a new billboard depicting the Oracle logo along with a broken samurai sword. The ad wasn't even really aimed at Oracle as an entity, let alone the consuming public; it was a personal attack on Ellison. But perhaps White spent a little too much time worrying about the competition: while he was busy creating billboards, Informix imploded in a massive accounting scandal and White soon found himself in federal prison for securities fraud.

If you can't beat a rival, it may be better to merge. I started Confinity with my co-founder Max Levchin in 1998. When we released the PayPal product in late 1999, Elon Musk's X.com was right on our heels: our companies' offices were four blocks apart on University Avenue in Palo Alto, and X's product mirrored ours feature-for-feature. By late 1999, we were in all-out war. Many of us at PayPal logged 100-hour workweeks. No doubt that was counterproductive, but the focus wasn't on objective productivity; the focus was defeating X.com. One of our engineers actually designed a bomb for this purpose; when he presented the schematic at a team meeting, calmer heads prevailed and the proposal was attributed to extreme sleep deprivation.

But in February 2000, Elon and I were more scared about the rapidly inflating tech bubble than we were about each other: a financial crash would ruin us both before we could finish our fight. So in early March we met on neutral ground—a café almost exactly equidistant to our offices— and negotiated a 50-50 merger. De-escalating the rivalry post-merger wasn't easy, but as far as problems go, it was a good one to have. As a unified team, we were able to ride out the dot-com crash and then build a successful business.

Sometimes you do have to fight. Where that's true, you should fight and win. There is no middle ground: either don't throw any punches, or strike hard and end it quickly.

This advice can be hard to follow because pride and honor can get in the way. Hence Hamlet:

> *Exposing what is mortal and unsure*
> *To all that fortune, death, and danger dare,*
> *Even for an eggshell. Rightly to be great*
> *Is not to stir without great argument,*
> *But greatly to find quarrel in a straw*
> *When honor's at the stake.*

For Hamlet, greatness means willingness to fight for reasons as thin as an eggshell: *anyone* would fight for things that matter; true heroes take their personal honor so seriously they will fight for things that *don't* matter. This twisted logic is part of human nature, but it's disastrous in business. If you can recognize competition as a destructive force instead of a sign of value, you're already more sane than most. The next chapter is about how to use a clear head to build a monopoly business.

# 5

# LAST MOVER
ADVANTAGE

Escaping competition will give you a monopoly, but even a monopoly is only a great business if it can endure in the future. Compare the value of the New York Times Company with Twitter. Each employs a few thousand people, and each gives millions of people a way to get news. But when Twitter went public in 2013, it was valued at $24 billion—*more than 12 times* the Times's market capitalization—even though the Times earned $133 million in 2012 while Twitter *lost* money. What explains the huge premium for Twitter?

The answer is cash flow. This sounds bizarre at first, since the Times was profitable while Twitter wasn't. But a great business is defined by its ability to generate cash flows *in the future*. Investors expect Twitter will be able to capture monopoly profits over the next decade, while newspapers' monopoly days are over.

Simply stated, the value of a business today is the sum of all the money it will make in the future. (To properly value a business, you also have to discount those future cash flows to their present worth, since a given amount of money today is worth more than the same amount in the future.)

Comparing discounted cash flows shows the difference between low-growth businesses and high-growth startups at its starkest. Most of the value of low-growth businesses is in the near term. An Old Economy business (like a newspaper) might hold its value if it can maintain its current cash flows for five or six years. However, any firm with close substitutes will see its profits competed away. Nightclubs or restaurants are extreme examples: successful ones might collect healthy amounts today, but their cash flows will probably dwindle over the next few years when customers move on to newer and trendier alternatives.

Technology companies follow the opposite trajectory. They often *lose* money for the first few years: it takes time to build valuable things, and that means delayed revenue. Most

## PRESENT VALUE CASH FLOWS OF A BUSINESS IN DECLINE

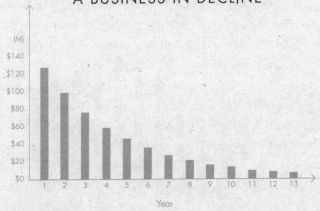

of a tech company's value will come at least 10 to 15 years in the future.

In March 2001, PayPal had yet to make a profit but our revenues were growing 100% year-over-year. When I projected our future cash flows, I found that 75% of the company's present value would come from profits generated in 2011 and beyond—hard to believe for a company that had been in business for only 27 months. But even that turned out to be an underestimation. Today, PayPal continues to grow at about 15% annually, and the discount rate is lower than a decade ago. It now appears that most of the company's value will come from 2020 and beyond.

LinkedIn is another good example of a company whose value exists in the far future. As of early 2014, its market capitalization was $24.5 billion—very high for a company with less than $1 billion in revenue and only $21.6 million in net income for 2012. You might look at these numbers and

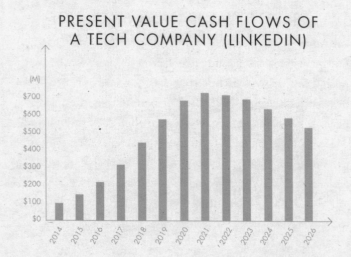

## PRESENT VALUE CASH FLOWS OF A TECH COMPANY (LINKEDIN)

conclude that investors have gone insane. But this valuation makes sense when you consider LinkedIn's projected future cash flows.

The overwhelming importance of future profits is counterintuitive even in Silicon Valley. For a company to be valuable it must grow *and endure,* but many entrepreneurs focus only on short-term growth. They have an excuse: growth is easy to measure, but durability isn't. Those who succumb to measurement mania obsess about weekly active user statistics, monthly revenue targets, and quarterly earnings reports. However, you can hit those numbers and still overlook deeper, harder-to-measure problems that threaten the durability of your business.

For example, rapid short-term growth at both Zynga and Groupon distracted managers and investors from long-term challenges. Zynga scored early wins with games like Farmville and claimed to have a "psychometric engine" to rigorously gauge the appeal of new releases. But they ended up with the same problem as every Hollywood studio: how can you reliably produce a constant stream of popular entertainment for a fickle audience? (Nobody knows.) Groupon posted fast growth as hundreds of thousands of local businesses tried their product. But persuading those businesses to become repeat customers was harder than they thought.

If you focus on near-term growth above all else, you miss the most important question you should be asking: *will this business still be around a decade from now?* Numbers alone won't tell you the answer; instead you must think critically about the qualitative characteristics of your business.

## CHARACTERISTICS OF MONOPOLY

What does a company with large cash flows far into the future look like? Every monopoly is unique, but they usually share some combination of the following characteristics: proprietary technology, network effects, economies of scale, and branding.

This isn't a list of boxes to check as you build your business—there's no shortcut to monopoly. However, analyzing your business according to these characteristics can help you think about how to make it durable.

### 1. Proprietary Technology

Proprietary technology is the most substantive advantage a company can have because it makes your product difficult or impossible to replicate. Google's search algorithms, for example, return results better than anyone else's. Proprietary technologies for extremely short page load times and highly accurate query autocompletion add to the core search product's robustness and defensibility. It would be very hard for anyone to do to Google what Google did to all the other search engine companies in the early 2000s.

As a good rule of thumb, proprietary technology must be at least 10 times better than its closest substitute in some important dimension to lead to a real monopolistic advantage. Anything less than an order of magnitude better will probably be perceived as a marginal improvement and will be hard to sell, especially in an already crowded market.

The clearest way to make a 10x improvement is to invent something completely new. If you build something valuable where there was nothing before, the increase in value is theoretically infinite. A drug to safely eliminate the need for sleep, or a cure for baldness, for example, would certainly support a monopoly business.

Or you can radically improve an existing solution: once you're 10x better, you escape competition. PayPal, for instance, made buying and selling on eBay at least 10 times better. Instead of mailing a check that would take 7 to 10 days to arrive, PayPal let buyers pay as soon as an auction ended. Sellers received their proceeds right away, and unlike with a check, they knew the funds were good.

Amazon made its first 10x improvement in a particularly visible way: they offered at least 10 times as many books as any other bookstore. When it launched in 1995, Amazon could claim to be "Earth's largest bookstore" because, unlike a retail bookstore that might stock 100,000 books, Amazon didn't need to physically store any inventory—it simply requested the title from its supplier whenever a customer made an order. This quantum improvement was so effective that a very unhappy Barnes & Noble filed a lawsuit three days before Amazon's IPO, claiming that Amazon was unfairly calling itself a "bookstore" when really it was a "book broker."

You can also make a 10x improvement through superior integrated design. Before 2010, tablet computing was so poor that for all practical purposes the market didn't even exist. "Microsoft Windows XP Tablet PC Edition" products first shipped in 2002, and Nokia released its own "Internet Tablet" in 2005, but they were a pain to use. Then Apple released the iPad. Design improvements are hard to measure, but it seems clear that Apple improved on anything that had come before by at least an order of magnitude: tablets went from unusable to useful.

## 2. Network Effects

Network effects make a product more useful as more people use it. For example, if all your friends are on Facebook, it makes sense for you to join Facebook, too. Unilaterally choosing a different social network would only make you an eccentric.

Network effects can be powerful, but you'll never reap them unless your product is valuable to its very first users when the network is necessarily small. For example, in 1960 a quixotic company called Xanadu set out to build a two-way communication network between all computers—a sort of early, synchronous version of the World Wide Web. After more than three decades of futile effort, Xanadu folded just as the web was becoming commonplace. Their technology probably would have worked at scale, but it could have worked *only* at scale: it required every computer to join the network at the same time, and that was never going to happen.

Paradoxically, then, network effects businesses must start with especially small markets. Facebook started with just Harvard students—Mark Zuckerberg's first product was designed to get all his classmates signed up, not to attract all people of Earth. This is why successful network businesses rarely get started by MBA types: the initial markets are so small that they often don't even appear to be business opportunities at all.

### *3. Economies of Scale*

A monopoly business gets stronger as it gets bigger: the fixed costs of creating a product (engineering, management, office space) can be spread out over ever greater quantities of sales. Software startups can enjoy especially dramatic economies of scale because the marginal cost of producing another copy of the product is close to zero.

Many businesses gain only limited advantages as they grow to large scale. Service businesses especially are difficult to make monopolies. If you own a yoga studio, for example, you'll only be able to serve a certain number of customers. You can hire more instructors and expand to more locations, but your margins will remain fairly low and you'll never reach a point where a core group of talented people can provide something of value to millions of separate clients, as software engineers are able to do.

A good startup should have the potential for great scale built into its first design. Twitter already has more than 250 million users today. It doesn't need to add too many customized features in order to acquire more, and there's no inherent reason why it should ever stop growing.

### *4. Branding*

A company has a monopoly on its own brand by definition, so creating a strong brand is a powerful way to claim a monopoly. Today's strongest tech brand is Apple: the attractive looks and carefully chosen materials of products like the iPhone and MacBook, the Apple Stores' sleek minimalist design and close control over the consumer experience, the omnipresent advertising campaigns, the price positioning as a maker of premium goods, and the lingering nimbus of Steve Jobs's personal charisma all contribute to a perception that Apple offers products so good as to constitute a category of their own.

Many have tried to learn from Apple's success: paid advertising, branded stores, luxurious materials, playful keynote speeches, high prices, and even minimalist design are all susceptible to imitation. But these techniques for polishing the surface don't work without a strong underlying substance. Apple has a complex suite of proprietary technologies, both in hardware (like superior touchscreen materials) and software (like touchscreen interfaces purpose-designed for specific materials). It manufactures products at a scale large enough to dominate pricing for the materials it buys. And it enjoys strong network effects from its content ecosystem: thousands of developers write software for Apple devices because that's where hundreds of·millions of users are, and those users stay on the platform because it's where the apps are. These other monopolistic advantages are less obvious than Apple's sparkling brand, but they are the fundamentals that let the branding effectively reinforce Apple's monopoly.

Beginning with brand rather than substance is dangerous. Ever since Marissa Mayer became CEO of Yahoo! in mid-2012, she has worked to revive the once-popular internet giant by making it cool again. In a single tweet, Yahoo! summarized Mayer's plan as a chain reaction of "people then products then traffic then revenue." The people are supposed to come for the coolness: Yahoo! demonstrated design awareness by overhauling its logo, it asserted youthful relevance by acquiring hot startups like Tumblr, and it has gained media attention for Mayer's own star power. But the big question is what products Yahoo! will actually create. When Steve Jobs returned to Apple, he didn't just make Apple a cool place to work; he slashed product lines to focus on the handful of opportunities for 10x improvements. No technology company can be built on branding alone.

## BUILDING A MONOPOLY

Brand, scale, network effects, and technology in some combination define a monopoly; but to get them to work, you need to choose your market carefully and expand deliberately.

### *Start Small and Monopolize*

Every startup is small at the start. Every monopoly dominates a large share of its market. *Therefore, every startup should start with a very small market.* Always err on the side of starting too small. The reason is simple: it's easier to dominate a small market than a large one. If you think your initial market might be too big, it almost certainly is.

Small doesn't mean nonexistent. We made this mistake early on at PayPal. Our first product let people beam money to each other via PalmPilots. It was interesting technology and no one else was doing it. However, the world's millions of PalmPilot users weren't concentrated in a particular place, they had little in common, and they used their devices only episodically. Nobody needed our product, so we had no customers.

With that lesson learned, we set our sights on eBay

auctions, where we found our first success. In late 1999, eBay had a few thousand high-volume "PowerSellers," and after only three months of dedicated effort, we were serving 25% of them. It was much easier to reach a few thousand people who really needed our product than to try to compete for the attention of millions of scattered individuals.

The perfect target market for a startup is a small group of particular people concentrated together and served by few or no competitors. Any big market is a bad choice, and a big market already served by competing companies is even worse. This is why it's always a red flag when entrepreneurs talk about getting 1% of a $100 billion market. In practice, a large market will either lack a good starting point or it will be open to competition, so it's hard to ever reach that 1%. And even if you do succeed in gaining a small foothold, you'll have to be satisfied with keeping the lights on: cut-throat competition means your profits will be zero.

## Scaling Up

Once you create and dominate a niche market, then you should gradually expand into related and slightly broader markets. Amazon shows how it can be done. Jeff Bezos's founding vision was to dominate all of online retail, but he very deliberately started with books. There were millions of books to catalog, but they all had roughly the same shape, they were easy to ship, and some of the most rarely sold books—those least profitable for any retail store to keep in stock—also drew the most enthusiastic customers. Amazon became the dominant solution for anyone located far from a bookstore or seeking something unusual. Amazon then had two options: expand the number of people who read books, or expand to adjacent markets. They chose the latter, starting with the most similar markets: CDs, videos, and software. Amazon continued to add categories gradually until it had become the world's general store. The name itself brilliantly encapsulated the company's scaling strategy. The biodiversity of the Amazon rain forest reflected Amazon's first goal of cataloging every book in the world, and now it stands for every kind of thing in the world, period.

eBay also started by dominating small niche markets. When it launched its auction marketplace in 1995, it didn't need the whole world to adopt it at once; the product worked well for intense interest groups, like Beanie Baby obsessives. Once it monopolized the Beanie Baby trade, eBay didn't jump straight to listing sports cars or industrial surplus: it continued to cater to small-time hobbyists until it became the most reliable marketplace for people trading online no matter what the item.

Sometimes there are hidden obstacles to scaling—a lesson that eBay has learned in recent years. Like all marketplaces, the auction marketplace lent itself to natural monopoly because buyers go where the sellers are and vice versa. But eBay found that the auction model works best for individually distinctive products like coins and stamps. It works less well for commodity products: people don't want to bid on pencils or Kleenex, so it's more convenient just to buy them from Amazon. eBay is still a valuable monopoly; it's just smaller than people in 2004 expected it to be.

Sequencing markets correctly is underrated, and it takes discipline to expand gradually. The most successful companies make the core progression—to first dominate a specific niche and then scale to adjacent markets—a part of their founding narrative.

## Don't Disrupt

Silicon Valley has become obsessed with "disruption." Originally, "disruption" was a term of art to describe how a firm can use new technology to introduce a low-end product at low prices, improve the product over time, and eventually overtake even the premium products offered by incumbent companies using older technology. This is roughly what happened when the advent of PCs disrupted the market for mainframe computers: at first PCs seemed irrelevant, then they became dominant. Today mobile devices may be doing the same thing to PCs.

However, disruption has recently transmogrified into a self-congratulatory buzzword for anything posing as trendy and new. This seemingly trivial fad matters because it distorts an entrepreneur's self-understanding in an inherently competitive way. The concept was coined to describe threats to incumbent companies, so startups' obsession with disruption means they see themselves through older firms' eyes. If you think of yourself as an insurgent battling dark forces, it's easy to become unduly fixated on the obstacles in your path. But if you truly want to make something new, the act of creation is far more important than the old industries that might not like what you create. Indeed, if your company can be summed up by its opposition to already existing firms, it can't be completely new and it's probably not going to become a monopoly.

Disruption also attracts attention: disruptors are people who look for trouble and find it. Disruptive kids get sent to the principal's office. Disruptive companies often pick fights they can't win. Think of Napster: the name itself meant trouble. What kinds of things can one "nap"? Music . . . Kids . . . and perhaps not much else. Shawn Fanning and Sean Parker, Napster's then-teenage founders, credibly threatened to disrupt the powerful music recording industry in 1999. The next year, they made the cover of *Time* magazine. A year and a half after that, they ended up in bankruptcy court.

PayPal could be seen as disruptive, but we didn't try to directly challenge any large competitor. It's true that we took some business away from Visa when we popularized internet payments: you might use PayPal to buy something online instead of using your Visa card to buy it in a store. But since we expanded the market for payments overall, we gave Visa far more business than we took. The overall dynamic was net positive, unlike Napster's negative-sum struggle with the U.S. recording industry. As you craft a plan to expand to adjacent markets, don't disrupt: avoid competition as much as possible.

## THE LAST WILL BE FIRST

You've probably heard about "first mover advantage": if you're the first entrant into a market, you can capture significant market share while competitors scramble to get started. But moving first is a tactic, not a goal. What really matters is generating cash flows in the future, so being the first mover doesn't do you any good if someone else comes along and unseats you. It's much better to be the *last* mover—that is, to make the last great development in a specific market and enjoy years or even decades of monopoly profits. The way to do that is to dominate a small niche and scale up from there, toward your ambitious long-term vision. In this one particular at least, business is like chess. Grandmaster José Raúl Capablanca put it well: to succeed, "you must study the endgame before everything else."

# 6

## YOU ARE NOT A LOTTERY TICKET

THE MOST CONTENTIOUS question in business is whether success comes from luck or skill.

What do successful people say? Malcolm Gladwell, a successful author who writes about successful people, declares in *Outliers* that success results from a "patchwork of lucky breaks and arbitrary advantages." Warren Buffett famously considers himself a "member of the lucky sperm club" and a winner of the "ovarian lottery." Jeff Bezos attributes Amazon's success to an "incredible planetary alignment" and jokes that it was "half luck, half good timing, and the rest brains." Bill Gates even goes so far as to claim that he "was lucky to be born with certain skills," though it's not clear whether that's actually possible.

Perhaps these guys are being strategically humble. However, the phenomenon of serial entrepreneurship would seem to call into question our tendency to explain success as the product of chance. Hundreds of people have started multiple multimillion-dollar businesses. A few, like Steve Jobs, Jack Dorsey, and Elon Musk, have created several multi*billion*-dollar companies. If success were mostly a matter of luck, these kinds of serial entrepreneurs probably wouldn't exist.

In January 2013, Jack Dorsey, founder of Twitter and Square, tweeted to his 2 million followers: "Success is never accidental."

Most of the replies were unambiguously negative. Referencing the tweet in *The Atlantic,* reporter Alexis Madrigal wrote that his instinct was to reply: "'Success is never accidental,' said all multimillionaire white men." It's true that already successful people have an easier time doing new things, whether due to their networks, wealth, or experience. But perhaps we've become too quick to dismiss anyone who claims to have succeeded according to plan.

Is there a way to settle this debate objectively? Unfortunately not, because companies are not experiments. To get a scientific answer about Facebook, for example, we'd have to rewind to 2004, create 1,000 copies of the world, and start Facebook in each copy to see how many times it would succeed. But that experiment is impossible. Every company starts in unique circumstances, and every company starts only once. Statistics doesn't work when the sample size is one.

From the Renaissance and the Enlightenment to the mid-20th century, luck was something to be mastered, dominated, and controlled; everyone agreed that you should do what you could, not focus on what you couldn't. Ralph Waldo Emerson captured this ethos when he wrote: "Shallow men believe in luck, believe in circumstances. . . . Strong men believe in cause and effect." In 1912, after he became the first explorer to reach the South Pole, Roald Amundsen wrote: "Victory awaits him who has everything in order— luck, people call it." No one pretended that misfortune didn't exist, but prior generations believed in making their own luck by working hard.

If you believe your life is mainly a matter of chance, why read this book? Learning about startups is worthless if you're just reading stories about people who won the lottery. *Slot Machines for Dummies* can purport to tell you which kind of rabbit's foot to rub or how to tell which machines are "hot," but it can't tell you how to win.

Did Bill Gates simply win the intelligence lottery? Was Sheryl Sandberg born with a silver spoon, or did she "lean in"? When we debate historical questions like these, luck is in the past tense. Far more important are questions about the future: is it a matter of chance or design?

## CAN YOU CONTROL YOUR FUTURE?

You can expect the future to take a definite form or you can treat it as hazily uncertain. If you treat the future as something definite, it makes sense to understand it in advance and to work to shape it. But if you expect an indefinite future ruled by randomness, you'll give up on trying to master it.

Indefinite attitudes to the future explain what's most dysfunctional in our world today. Process trumps substance: when people lack concrete plans to carry out, they use formal rules to assemble a portfolio of various options. This describes

Americans today. In middle school, we're encouraged to start hoarding "extracurricular activities." In high school, ambitious students compete even harder to appear omnicompetent. By the time a student gets to college, he's spent a decade curating a bewilderingly diverse résumé to prepare for a completely unknowable future. Come what may, he's ready—for nothing in particular.

A definite view, by contrast, favors firm convictions. Instead of pursuing many-sided mediocrity and calling it "well-roundedness," a definite person determines the one best thing to do and then does it. Instead of working tirelessly to make herself indistinguishable, she strives to be great at something substantive—to be a monopoly of one. This is not what young people do today, because everyone around them has long since lost faith in a definite world. No one gets into Stanford by excelling at just one thing, unless that thing happens to involve throwing or catching a leather ball.

|  | DEFINITE | INDEFINITE |
|---|---|---|
| OPTIMISTIC | U.S., 1950s–1960s | U.S., 1982–present |
| PESSIMISTIC | China, present | Europe, present |

You can also expect the future to be either better or worse than the present. Optimists welcome the future; pessimists fear it. Combining these possibilities yields four views:

## Indefinite Pessimism

Every culture has a myth of decline from some golden age, and almost all peoples throughout history have been pessimists. Even today pessimism still dominates huge parts of the world. An *indefinite pessimist* looks out onto a bleak future, but he has no idea what to do about it. This describes Europe since the early 1970s, when the continent succumbed to undirected bureaucratic drift. Today the whole Eurozone is in slow-motion crisis, and nobody is in charge. The European Central Bank doesn't stand for anything but improvisation: the U.S. Treasury prints "In God We Trust" on the dollar; the ECB might as well print "Kick the Can Down the Road" on the euro. Europeans just react to events as they happen and hope things don't get worse. The indefinite pessimist can't know whether the inevitable decline will be fast or slow, catastrophic or gradual. All he can do is wait for it to happen, so he might as well eat, drink, and be merry in the meantime: hence Europe's famous vacation mania.

## Definite Pessimism

A *definite pessimist* believes the future can be known, but since it will be bleak, he must prepare for it. Perhaps surprisingly, China is probably the most definitely pessimistic place in the world today. When Americans see the Chinese economy grow ferociously fast (10% per year since 2000), we imagine a confident country mastering its future. But that's because

Americans are still optimists, and we project our optimism onto China. From China's viewpoint, economic growth cannot come fast enough. Every other country is afraid that China is going to take over the world; China is the only country afraid that it won't.

China can grow so fast only because its starting base is so low. The easiest way for China to grow is to relentlessly copy what has already worked in the West. And that's exactly what it's doing: executing definite plans by burning ever more coal to build ever more factories and skyscrapers. But with a huge population pushing resource prices higher, there's no way Chinese living standards can ever actually catch up to those of the richest countries, and the Chinese know it.

This is why the Chinese leadership is obsessed with the way in which things threaten to get worse. Every senior Chinese leader experienced famine as a child, so when the Politburo looks to the future, disaster is not an abstraction. The Chinese public, too, knows that winter is coming. Outsiders are fascinated by the great fortunes being made inside China, but they pay less attention to the wealthy Chinese trying hard to get their money out of the country. Poorer Chinese just save everything they can and hope it will be enough. Every class of people in China takes the future deadly seriously.

## Definite Optimism

To a *definite optimist,* the future will be better than the present if he plans and works to make it better. From the 17th century through the 1950s and '60s, definite optimists led the Western world. Scientists, engineers, doctors, and businessmen made the world richer, healthier, and more long-lived than previously imaginable. As Karl Marx and Friedrich Engels saw clearly, the 19th-century business class

> created more massive and more colossal productive
> forces than all preceding generations together.
> Subjection of Nature's forces to man, machinery,
> application of chemistry to industry and agriculture,
> steam-navigation, railways, electric telegraphs,
> clearing of whole continents for cultivation, canalisation
> of rivers, whole populations conjured out of the
> ground—what earlier century had even a presentiment
> that such productive forces slumbered in the lap of
> social labor?

Each generation's inventors and visionaries surpassed their predecessors. In 1843, the London public was invited to make its first crossing underneath the River Thames by a newly dug tunnel. In 1869, the Suez Canal saved Eurasian shipping traffic from rounding the Cape of Good Hope. In 1914 the Panama Canal cut short the route from Atlantic to Pacific. Even the Great Depression failed to impede relentless progress in the United States, which has always been home to the world's most far-seeing definite optimists. The Empire State Building was started in 1929 and finished in 1931. The Golden Gate Bridge was started in 1933 and completed in 1937. The Manhattan Project was started in 1941 and had already produced the world's first nuclear bomb by 1945. Americans continued to remake the face of the world in peacetime: the Interstate Highway System began construction in 1956, and the first 20,000 miles of road were open for driving by 1965. Definite planning even went beyond the surface of this planet: NASA's Apollo Program began in 1961 and put 12 men on the moon before it finished in 1972.

Bold plans were not reserved just for political leaders or government scientists. In the late 1940s, a Californian named John Reber set out to reinvent the physical geography of the whole San Francisco Bay Area. Reber was a schoolteacher, an amateur theater producer, and a self-taught engineer. Undaunted by his lack of credentials, he publicly proposed to build two huge dams in the Bay, construct massive freshwater lakes for drinking water and irrigation, and reclaim 20,000 acres of land for development. Even though he had no personal authority, people took the Reber Plan seriously. It was endorsed by newspaper editorial boards across California. The U.S. Congress held hearings on its feasibility. The Army Corps of Engineers even constructed a 1.5-acre scale model of the Bay in a cavernous Sausalito warehouse to simulate it. These tests revealed technical shortcomings, so the plan wasn't executed.

But would anybody today take such a vision seriously in the first place? In the 1950s, people welcomed big plans and asked whether they would work. Today a grand plan coming from a schoolteacher would be dismissed as crankery, and a long-range vision coming from anyone more powerful would be derided as hubris. You can still visit the Bay Model in that Sausalito warehouse, but today it's just a tourist attraction: big plans for the future have become archaic curiosities.

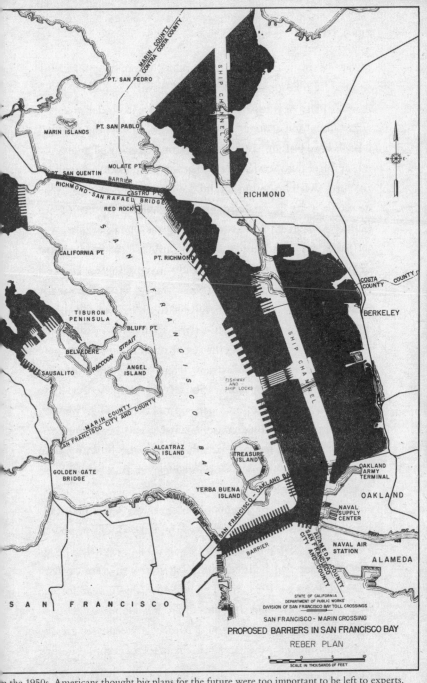

STATE OF CALIFORNIA
DEPARTMENT OF PUBLIC WORKS
DIVISION OF SAN FRANCISCO BAY TOLL CROSSINGS

SAN FRANCISCO - MARIN CROSSING

**PROPOSED BARRIERS IN SAN FRANCISCO BAY**

REBER PLAN

SCALE IN THOUSANDS OF FEET

n the 1950s, Americans thought big plans for the future were too important to be left to experts.

## Indefinite Optimism

After a brief pessimistic phase in the 1970s, indefinite optimism has dominated American thinking ever since 1982, when a long bull market began and finance eclipsed engineering as the way to approach the future. To an *indefinite optimist,* the future will be better, but he doesn't know how exactly, so he won't make any specific plans. He expects to profit from the future but sees no reason to design it concretely.

Instead of working for years to build a new product, indefinite optimists rearrange already-invented ones. Bankers make money by rearranging the capital structures of already existing companies. Lawyers resolve disputes over old things or help other people structure their affairs. And private equity investors and management consultants don't start new businesses; they squeeze extra efficiency from old ones with incessant procedural optimizations. It's no surprise that these fields all attract disproportionate numbers of high-achieving Ivy League optionality chasers; what could be a more appropriate reward for two decades of résumé-building than a seemingly elite, process-oriented career that promises to "keep options open"?

Recent graduates' parents often cheer them on the established path. The strange history of the Baby Boom produced a generation of indefinite optimists so used to effortless progress that they feel entitled to it. Whether you were born in 1945 or 1950 or 1955, things got better every year for the first 18 years of your life, *and it had nothing to do with you.* Technological advance seemed to accelerate automatically, so the Boomers grew up with great expectations but few specific plans for how to fulfill them. Then, when technological progress stalled in the 1970s, increasing income inequality came to the rescue of the most elite Boomers. Every year of adulthood continued to get automatically better and better for the rich and successful. The rest of their generation was left behind, but the wealthy Boomers who shape public opinion today see little reason to question their naïve optimism. Since tracked careers worked for them, they can't imagine that they won't work for their kids, too.

Malcolm Gladwell says you can't understand Bill Gates's success without understanding his fortunate personal context: he grew up in a good family, went to a private school equipped with a computer lab, and counted Paul Allen as a childhood friend. But perhaps you can't understand Malcolm Gladwell without understanding *his* historical context as a Boomer (born in 1963). When Baby Boomers grow up and write books to explain why one or another individual is successful, they point to the power of a particular individual's context as determined by chance. But they miss the even bigger social context for their own preferred explanations: a whole generation learned from childhood to overrate the power of chance and underrate the importance of planning. Gladwell at first appears to be making a contrarian critique of the myth of the self-made businessman, but actually his own account encapsulates the conventional view of a generation.

# OUR INDEFINITELY OPTIMISTIC WORLD

## *Indefinite Finance*

While a definitely optimistic future would need engineers to design underwater cities and settlements in space, an indefinitely optimistic future calls for more bankers and lawyers. Finance epitomizes indefinite thinking because it's the only way to make money when you have no idea how to create wealth. If they don't go to law school, bright college graduates head to Wall Street precisely because they have no real plan for their careers. And once they arrive at Goldman, they find that even *inside* finance, everything is indefinite. It's still optimistic—you wouldn't play in the markets if you expected to lose—but the fundamental tenet is that the market is random; you can't know anything specific or substantive; diversification becomes supremely important.

The indefiniteness of finance can be bizarre. Think about what happens when successful entrepreneurs sell their company. What do they do with the money? In a financialized world, it unfolds like this:

- The founders don't know what to do with it, so they give it to a large bank.

- The bankers don't know what to do with it, so they diversify by spreading it across a portfolio of institutional investors.

- Institutional investors don't know what to do with their managed capital, so they diversify by amassing a portfolio of stocks.

- Companies try to increase their share price by
  generating free cash flows. If they do, they issue
  dividends or buy back shares and the cycle repeats.

At no point does anyone in the chain know what to do
with money in the real economy. But in an indefinite world,
people actually *prefer* unlimited optionality; money is more
valuable than anything you could possibly do with it. Only
in a definite future is money a means to an end, not the end
itself.

## Indefinite Politics

Politicians have always been officially accountable to the public
at election time, but today they are attuned to what the public
thinks *at every moment*. Modern polling enables politicians to
tailor their image to match preexisting public opinion exactly,
so for the most part, they do. Nate Silver's election predictions
are remarkably accurate, but even more remarkable is how big
a story they become every four years. We are more fascinated
today by statistical predictions of what the country will be
thinking in a few weeks' time than by visionary predictions
of what the country will look like 10 or 20 years from now.

And it's not just the electoral process—the very character
of government has become indefinite, too. The government
used to be able to coordinate complex solutions to problems
like atomic weaponry and lunar exploration. But today, after
40 years of indefinite creep, the government mainly just
provides insurance; our solutions to big problems are Medi-
care, Social Security, and a dizzying array of other transfer

payment programs. It's no surprise that entitlement spending has eclipsed discretionary spending every year since 1975. To increase discretionary spending we'd need definite plans to solve specific problems. But according to the indefinite logic of entitlement spending, we can make things better just by sending out more checks.

## Indefinite Philosophy

You can see the shift to an indefinite attitude not just in politics but in the political philosophers whose ideas underpin both left and right.

The philosophy of the ancient world was pessimistic: Plato, Aristotle, Epicurus, and Lucretius all accepted strict limits on human potential. The only question was how best to cope with our tragic fate. Modern philosophers have been mostly optimistic. From Herbert Spencer on the right and Hegel in the center to Marx on the left, the 19th century shared a belief in progress. (Remember Marx and Engels's encomium to the technological triumphs of capitalism from page 65.) These thinkers expected material advances to fundamentally change human life for the better: they were definite optimists.

In the late 20th century, indefinite philosophies came to the fore. The two dominant political thinkers, John Rawls and Robert Nozick, are usually seen as stark opposites: on the egalitarian left, Rawls was concerned with questions of fairness and distribution; on the libertarian right, Nozick focused on maximizing individual freedom. They both believed that people could get along with each other peacefully, so unlike the ancients, they were optimistic. But unlike

Spencer or Marx, Rawls and Nozick were *indefinite* optimists: they didn't have any specific vision of the future.

|  | DEFINITE | INDEFINITE |
|---|---|---|
| OPTIMISTIC | Hegel, Marx | Nozick, Rawls |
| PESSIMISTIC | Plato, Aristotle | Epicurus, Lucretius |

Their indefiniteness took different forms. Rawls begins *A Theory of Justice* with the famous "veil of ignorance": fair political reasoning is supposed to be impossible for anyone with knowledge of the world as it concretely exists. Instead of trying to change our actual world of unique people and real technologies, Rawls fantasized about an "inherently stable" society with lots of fairness but little dynamism. Nozick opposed Rawls's "patterned" concept of justice. To Nozick, any voluntary exchange must be allowed, and no social pattern could be noble enough to justify maintenance by coercion. He didn't have any more concrete ideas about the good society than Rawls: both of them focused on process. Today, we exaggerate the differences between left-liberal egalitarianism and libertarian individualism because almost everyone shares their common indefinite attitude. In philosophy, politics, and business, too, arguing over process has become a way to endlessly defer making concrete plans for a better future.

## Indefinite Life

Our ancestors sought to understand and extend the human lifespan. In the 16th century, conquistadors searched the jungles of Florida for a Fountain of Youth. Francis Bacon wrote that "the prolongation of life" should be considered its own branch of medicine—and the noblest. In the 1660s, Robert Boyle placed life extension (along with "the Recovery of Youth") atop his famous wish list for the future of science. Whether through geographic exploration or laboratory research, the best minds of the Renaissance thought of death as something to defeat. (Some resisters were killed in action: Bacon caught pneumonia and died in 1626 while experimenting to see if he could extend a chicken's life by freezing it in the snow.)

We haven't yet uncovered the secrets of life, but insurers and statisticians in the 19th century successfully revealed a secret about death that still governs our thinking today: they discovered how to reduce it to a mathematical probability. "Life tables" tell us our chances of dying in any given year, something previous generations didn't know. However, in exchange for better insurance contracts, we seem to have given up the search for secrets about longevity. Systematic knowledge of the current range of human lifespans has made that range seem natural. Today our society is permeated by the twin ideas that death is both inevitable and random.

Meanwhile, probabilistic attitudes have come to shape the agenda of biology itself. In 1928, Scottish scientist Alexander

Fleming found that a mysterious antibacterial fungus had grown on a petri dish he'd forgotten to cover in his laboratory: he discovered penicillin by accident. Scientists have sought to harness the power of chance ever since. Modern drug discovery aims to amplify Fleming's serendipitous circumstances a millionfold: pharmaceutical companies search through combinations of molecular compounds at random, hoping to find a hit.

But it's not working as well as it used to. Despite dramatic advances over the past two centuries, in recent decades biotechnology hasn't met the expectations of investors—or patients. Eroom's law—that's Moore's law backward—observes that the number of new drugs approved per billion dollars spent on R&D has halved every nine years since 1950. Since information technology accelerated faster than ever during those same years, the big question for biotech today is whether it will ever see similar progress. Compare biotech startups to their counterparts in computer software:

|  | Biotech Startups | Software Startups |
|---|---|---|
| Subject | Uncontrollable organisms | Perfectly determinate code |
| Environment | Poorly understood, natural | Well understood, artificial |
| Approach | Indefinite, random | Definite, engineering |
| Regulation | Heavily regulated | Basically unregulated |
| Cost | Expensive ( > $1B per drug) | Cheap (a little seed money) |
| Team | High-salaried, unaligned lab drones | Committed entrepreneurial hackers |

Biotech startups are an extreme example of indefinite thinking. Researchers experiment with things that just might work instead of refining definite theories about how the body's systems operate. Biologists say they need to work this way because the underlying biology is hard. According to them, IT startups work because we created computers ourselves and designed them to reliably obey our commands. Biotech is difficult because we didn't design our bodies, and the more we learn about them, the more complex they turn out to be.

But today it's possible to wonder whether the genuine difficulty of biology has become an excuse for biotech startups' indefinite approach to business in general. Most of the people involved expect some things to work eventually, but few want to commit to a specific company with the level of intensity necessary for success. It starts with the professors who often become part-time consultants instead of full-time employees—even for the biotech startups that begin from their own research. Then everyone else imitates the professors' indefinite attitude. It's easy for libertarians to claim that heavy regulation holds biotech back—and it does—but indefinite optimism may pose an even greater challenge for the future of biotech.

## IS INDEFINITE OPTIMISM EVEN POSSIBLE?

What kind of future will our indefinitely optimistic decisions bring about? If American households were saving, at least they could expect to have money to spend later. And if

American companies were investing, they could expect to reap the rewards of new wealth in the future. But U.S. households are saving almost nothing. And U.S. companies are letting cash pile up on their balance sheets without investing in new projects because they don't have any concrete plans for the future.

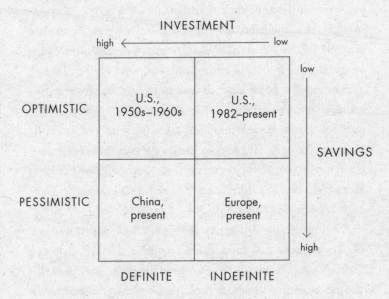

The other three views of the future can work. Definite optimism works when you build the future you envision. Definite pessimism works by building what can be copied without expecting anything new. Indefinite pessimism works because it's self-fulfilling: if you're a slacker with low expectations, they'll probably be met. But indefinite optimism seems inherently unsustainable: how can the future get better if no one plans for it?

Actually, most everybody in the modern world has already heard an answer to this question: progress without planning is what we call "evolution." Darwin himself wrote that life tends to "progress" without anybody intending it. Every living thing is just a random iteration on some other organism, and the best iterations win.

Darwin's theory explains the origin of trilobites and dinosaurs, but can it be extended to domains that are far removed? Just as Newtonian physics can't explain black holes or the Big Bang, it's not clear that Darwinian biology should explain how to build a better society or how to create a new business out of nothing. Yet in recent years Darwinian (or pseudo-Darwinian) metaphors have become common in business. Journalists analogize literal survival in competitive ecosystems to corporate survival in competitive markets. Hence all the headlines like "Digital Darwinism," "Dot-com Darwinism," and "Survival of the Clickiest."

Even in engineering-driven Silicon Valley, the buzzwords of the moment call for building a "lean startup" that can "adapt" and "evolve" to an ever-changing environment. Would-be entrepreneurs are told that nothing can be known in advance: we're supposed to listen to what customers say they want, make nothing more than a "minimum viable product," and iterate our way to success.

But leanness is a methodology, not a goal. Making small changes to things that already exist might lead you to a local maximum, but it won't help you find the global maximum. You could build the best version of an app that lets people order toilet paper from their iPhone. But iteration without a bold plan won't take you from 0 to 1. A company is the strangest place of all for an indefinite optimist: why should you expect your own business to succeed without a plan to make it happen? Darwinism may be a fine theory in other contexts, but in startups, intelligent design works best.

## THE RETURN OF DESIGN

What would it mean to prioritize design over chance? Today, "good design" is an aesthetic imperative, and everybody from slackers to yuppies carefully "curates" their outward appearance. It's true that every great entrepreneur is first and foremost a designer. Anyone who has held an iDevice or a smoothly machined MacBook has felt the result of Steve Jobs's obsession with visual and experiential perfection. But the most important lesson to learn from Jobs has nothing to do with aesthetics. The greatest thing Jobs designed was his business. Apple imagined and executed definite multi-year plans to create new products and distribute them effectively. Forget "minimum viable products"—ever since he started Apple in 1976, Jobs saw that you can change the world through careful planning, not by listening to focus group feedback or copying others' successes.

Long-term planning is often undervalued by our indefinite short-term world. When the first iPod was released in October 2001, industry analysts couldn't see much more than "a nice feature for Macintosh users" that "doesn't make any difference" to the rest of the world. Jobs planned the iPod to be the first of a new generation of portable post-PC devices, but that secret was invisible to most people. One look at the company's stock chart shows the harvest of this multi-year plan:

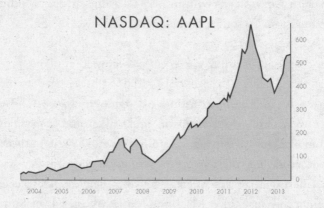

NASDAQ: AAPL

The power of planning explains the difficulty of valuing private companies. When a big company makes an offer to acquire a successful startup, it almost always offers too much or too little: founders only sell when they have no more concrete visions for the company, in which case the acquirer probably overpaid; definite founders with robust plans don't sell, which means the offer wasn't high enough. When Yahoo! offered to buy Facebook for $1 billion in July 2006, I thought we should at least consider it. But Mark Zuckerberg walked into the board meeting and announced: "Okay, guys, this is just a formality, it shouldn't take more than 10 minutes. We're obviously not going to sell here." Mark saw where he could take the company, and Yahoo! didn't. A business with a good definite plan will always be underrated in a world where people see the future as random.

# YOU ARE NOT A LOTTERY TICKET

We have to find our way back to a definite future, and the Western world needs nothing short of a cultural revolution to do it.

Where to start? John Rawls will need to be displaced in philosophy departments. Malcolm Gladwell must be persuaded to change his theories. And pollsters have to be driven from politics. But the philosophy professors and the Gladwells of the world are set in their ways, to say nothing of our politicians. It's extremely hard to make changes in those crowded fields, even with brains and good intentions.

A startup is the largest endeavor over which you can have definite mastery. You can have agency not just over your own life, but over a small and important part of the world. It begins by rejecting the unjust tyranny of Chance. You are not a lottery ticket.

# 7

# FOLLOW THE MONEY

ONEY MAKES MONEY. "For whoever has will be given more, and they will have an abundance. Whoever does not have, even what they have will be taken from them" (Matthew 25:29). Albert Einstein made the same observation when he stated that compound interest was "the eighth wonder of the world," "the greatest mathematical discovery of all time," or even "the most powerful force in the universe." Whichever version you prefer, you can't miss his message: never underestimate exponential growth. Actually, there's no evidence that Einstein ever said any of those things—the quotations are all apocryphal. But this very misattribution reinforces the message: having invested the principal of a lifetime's brilliance, Einstein continues to earn interest on it from beyond the grave by receiving credit for things he never said.

Most sayings are forgotten. At the other extreme, a select few people like Einstein and Shakespeare are constantly quoted and ventriloquized. We shouldn't be surprised, since small minorities often achieve disproportionate results. In

1906, economist Vilfredo Pareto discovered what became the "Pareto principle," or the 80-20 rule, when he noticed that 20% of the people owned 80% of the land in Italy—a phenomenon that he found just as natural as the fact that 20% of the peapods in his garden produced 80% of the peas. This extraordinarily stark pattern, in which a small few radically outstrip all rivals, surrounds us everywhere in the natural and social world. The most destructive earthquakes are many times more powerful than all smaller earthquakes combined. The biggest cities dwarf all mere towns put together. And monopoly businesses capture more value than millions of undifferentiated competitors. Whatever Einstein did or didn't say, the power law—so named because exponential equations describe severely unequal distributions—is the law of the universe. It defines our surroundings so completely that we usually don't even see it.

This chapter shows how the power law becomes visible when you follow the money: in venture capital, where investors try to profit from exponential growth in early-stage companies, a few companies attain exponentially greater value than all others. Most businesses never need to deal with venture capital, but everyone needs to know exactly one thing that even venture capitalists struggle to understand: we don't live in a normal world; we live under a power law.

## THE POWER LAW OF VENTURE CAPITAL

Venture capitalists aim to identify, fund, and profit from promising early-stage companies. They raise money from institutions and wealthy people, pool it into a fund, and invest in technology companies that they believe will become more valuable. If they turn out to be right, they take a cut of the returns—usually 20%. A venture fund makes money when the companies in its portfolio become more valuable and either go public or get bought by larger companies. Venture funds usually have a 10-year lifespan since it takes time for successful companies to grow and "exit."

But most venture-backed companies don't IPO or get acquired; most fail, usually soon after they start. Due to these early failures, a venture fund typically loses money at first. VCs hope the value of the fund will increase dramatically in a few years' time, to break-even and beyond, when the successful portfolio companies hit their exponential growth spurts and start to scale.

The big question is when this takeoff will happen. For most funds, the answer is never. Most startups fail, and most funds fail with them. Every VC knows that his task is to find

## J-CURVE OF A SUCCESSFUL VENTURE FUND

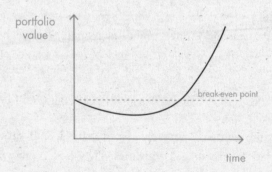

the companies that will succeed. However, even seasoned investors understand this phenomenon only superficially. They know companies are different, but they underestimate the degree of difference.

The error lies in expecting that venture returns will be normally distributed: that is, bad companies will fail, mediocre ones will stay flat, and good ones will return 2x or even 4x. Assuming this bland pattern, investors assemble a diversified portfolio and hope that winners counterbalance losers.

But this "spray and pray" approach usually produces an entire portfolio of flops, with no hits at all. This is because venture returns don't follow a normal distribution overall. Rather, they follow a power law: a small handful of companies radically outperform all others. If you focus on diversification instead of single-minded pursuit of the very few companies that can become overwhelmingly valuable, you'll miss those rare companies in the first place.

This graph shows the stark reality versus the perceived relative homogeneity:

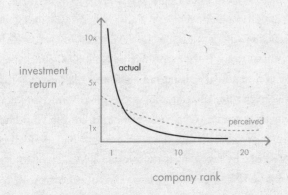

Our results at Founders Fund illustrate this skewed pattern: Facebook, the best investment in our 2005 fund, returned more than all the others combined. Palantir, the second-best investment, is set to return more than the sum of every other investment aside from Facebook. This highly uneven pattern is not unusual: we see it in all our other funds as well. *The biggest secret in venture capital is that the best investment in a successful fund equals or outperforms the entire rest of the fund combined.*

This implies two very strange rules for VCs. First, only invest in companies that have the potential to return the value of the entire fund. This is a scary rule, because it eliminates the vast majority of possible investments. (Even quite successful companies usually succeed on a more humble scale.) This leads to rule number two: because rule number one is so restrictive, there can't be any other rules.

Consider what happens when you break the first rule. Andreessen Horowitz invested $250,000 in Instagram in 2010. When Facebook bought Instagram just two years later for $1 billion, Andreessen netted $78 million—a 312x return in less than two years. That's a phenomenal return, befitting the firm's reputation as one of the Valley's best. But in a weird way it's not nearly enough, because Andreessen Horowitz has a $1.5 billion fund: if they only wrote $250,000 checks, they would need to find 19 Instagrams just to break even. This is why investors typically put a lot more money into any company worth funding. (And to be fair, Andreessen would have invested more in Instagram's later rounds had it not been conflicted out by a previous investment.) VCs must find the handful of companies that will successfully go from 0 to 1 and then back them with every resource.

Of course, no one can know with certainty *ex ante* which companies will succeed, so even the best VC firms have a "portfolio." However, *every single company in a good venture portfolio must have the potential to succeed at vast scale*. At Founders Fund, we focus on five to seven companies in a fund, each of which we think could become a multibillion-dollar business based on its unique fundamentals. Whenever you shift from the substance of a business to the financial question of whether or not it fits into a diversified hedging strategy, venture investing starts to look a lot like buying lottery tickets. And once you think that you're playing the lottery, you've already psychologically prepared yourself to lose.

## WHY PEOPLE DON'T SEE THE POWER LAW

Why would professional VCs, of all people, fail to see the power law? For one thing, it only becomes clear over time, and even technology investors too often live in the present.

## BEGINNING OF FUND

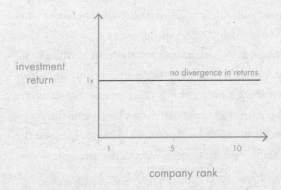

Imagine a firm invests in 10 companies with the potential to become monopolies—already an unusually disciplined portfolio. Those companies will look very similar in the early stages before exponential growth.

Over the next few years, some companies will fail while others begin to succeed; valuations will diverge, but the difference between exponential growth and linear growth will be unclear.

## MID–FUND

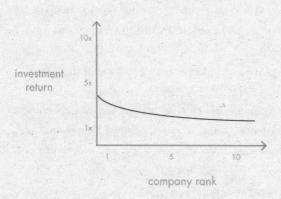

After 10 years, however, the portfolio won't be divided between winners and losers; it will be split between one dominant investment and everything else.

But no matter how unambiguous the end result of the power law, it doesn't reflect daily experience. Since investors spend most of their time making new investments and attending to companies in their early stages, most of the companies they work with are by definition average. Most of the differences that investors and entrepreneurs perceive every

## MATURE FUND

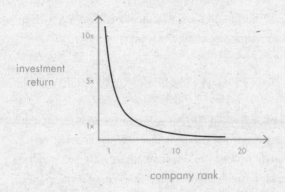

day are between relative levels of success, not between exponential dominance and failure. And since nobody wants to give up on an investment, VCs usually spend even more time on the most problematic companies than they do on the most obviously successful.

If even investors specializing in exponentially growing startups miss the power law, it's not surprising that most everyone else misses it, too. Power law distributions are so big that they hide in plain sight. For example, when most people outside Silicon Valley think of venture capital, they might picture a small and quirky coterie—like ABC's *Shark Tank,* only without commercials. After all, less than 1% of new businesses started each year in the U.S. receive venture funding, and total VC investment accounts for less than 0.2% of GDP. But the results of those investments disproportionately propel the entire economy. Venture-backed companies create 11% of all private sector jobs. They generate annual revenues equivalent to an astounding 21% of GDP. Indeed, the dozen largest tech companies were all venture-backed. Together those 12 companies are worth more than $2 trillion, *more than all other tech companies combined.*

## WHAT TO DO WITH THE POWER LAW

The power law is not just important to investors; rather, it's important to everybody because everybody is an investor. An entrepreneur makes a major investment just by spending her time working on a startup. Therefore every entrepreneur must think about whether her company is going to succeed and become valuable. Every individual is unavoidably an investor, too. When you choose a career, you act on your belief that the kind of work you do will be valuable decades from now.

The most common answer to the question of future value is a diversified portfolio: "Don't put all your eggs in one basket," everyone has been told. As we said, even the best venture investors have a portfolio, but investors who understand the power law make as few investments as possible. The kind of portfolio thinking embraced by both folk wisdom and financial convention, by contrast, regards diversified betting as a source of strength. The more you dabble, the more you are supposed to have hedged against the uncertainty of the future.

But life is not a portfolio: not for a startup founder, and not for any individual. An entrepreneur cannot "diversify" herself: you cannot run dozens of companies at the same time and then hope that one of them works out well. Less obvious but just as important, an individual cannot diversify his own life by keeping dozens of equally possible careers in ready reserve.

Our schools teach the opposite: institutionalized education traffics in a kind of homogenized, generic knowledge. Everybody who passes through the American school system learns *not* to think in power law terms. Every high school course period lasts 45 minutes whatever the subject. Every student proceeds at a similar pace. At college, model students obsessively hedge their futures by assembling a suite of exotic and minor skills. Every university believes in "excellence," and hundred-page course catalogs arranged alphabetically according to arbitrary departments of knowledge seem designed to reassure you that "it doesn't matter what you do, as long as you do it well." That is completely false. It does matter what you do. You should focus relentlessly on something you're good at doing, but before that you must think hard about whether it will be valuable in the future.

For the startup world, this means you should not necessarily start your own company, even if you are extraordinarily talented. If anything, too many people are starting their own companies today. People who understand the power law will hesitate more than others when it comes to founding a new venture: they know how tremendously successful they could become by joining the very best company while it's growing fast. The power law means that differences *between* companies will dwarf the differences in roles *inside* companies. You could have 100% of the equity if you fully fund your own venture, but if it fails you'll have 100% of nothing. Owning just 0.01% of Google, by contrast, is incredibly valuable (more than $35 million as of this writing).

If you do start your own company, you must remember the power law to operate it well. The most important things are singular: One market will probably be better than all others, as we discussed in Chapter 5. One distribution strategy usually dominates all others, too—for that see Chapter 11. Time and decision-making themselves follow a power law, and some moments matter far more than others—see Chapter 9. However, you can't trust a world that denies the power law to accurately frame your decisions for you, so what's most important is rarely obvious. It might even be secret. But in a power law world, you can't afford not to think hard about where your actions will fall on the curve.

# 8

## SECRETS

EVERY ONE OF TODAY'S most famous and familiar ideas was once unknown and unsuspected. The mathematical relationship between a triangle's sides, for example, was secret for millennia. Pythagoras had to think hard to discover it. If you wanted in on Pythagoras's new discovery, joining his strange vegetarian cult was the best way to learn about it. Today, his geometry has become a convention—a simple truth we teach to grade schoolers. A conventional truth can be important—it's essential to learn elementary mathematics, for example—but it won't give you an edge. It's not a secret.

Remember our contrarian question: *what important truth do very few people agree with you on?* If we already understand as much of the natural world as we ever will—if all of today's conventional ideas are already enlightened, and if everything

has already been done—then there are no good answers. Contrarian thinking doesn't make any sense unless the world still has secrets left to give up.

Of course, there are many things we don't yet understand, but some of those things may be impossible to figure out—mysteries rather than secrets. For example, string theory describes the physics of the universe in terms of vibrating one-dimensional objects called "strings." Is string theory true? You can't really design experiments to test it. Very few people, if any, could ever understand all its implications. But is that just because it's difficult? Or is it an impossible mystery? The difference matters. You can achieve difficult things, but you can't achieve the impossible.

Recall the business version of our contrarian question: *what valuable company is nobody building?* Every correct answer is necessarily a secret: something important and unknown, something hard to do but doable. If there are many secrets left in the world, there are probably many world-changing companies yet to be started. This chapter will help you think about secrets and how to find them.

## WHY AREN'T PEOPLE LOOKING FOR SECRETS?

Most people act as if there were no secrets left to find. An extreme representative of this view is Ted Kaczynski, infamously known as the Unabomber. Kaczynski was a child prodigy who enrolled at Harvard at 16. He went on to get a PhD in math and become a professor at UC Berkeley. But you've only ever heard of him because of the 17-year terror campaign he waged with pipe bombs against professors, technologists, and businesspeople.

In late 1995, the authorities didn't know who or where the Unabomber was. The biggest clue was a 35,000-word manifesto that Kaczynski had written and anonymously mailed to the press. The FBI asked some prominent newspapers to publish it, hoping for a break in the case. It worked: Kaczynski's brother recognized his writing style and turned him in.

You might expect that writing style to have shown obvious signs of insanity, but the manifesto is eerily cogent. Kaczynski claimed that in order to be happy, every individual "needs to have goals whose attainment requires effort, and needs to succeed in attaining at least some of his goals." He divided human goals into three groups:

1. Goals that can be satisfied with minimal effort;

2. Goals that can be satisfied with serious effort; and

3. Goals that cannot be satisfied, no matter how much effort one makes.

This is the classic trichotomy of the easy, the hard, and the impossible. Kaczynski argued that modern people are depressed because all the world's hard problems have already been solved. What's left to do is either easy or impossible, and pursuing those tasks is deeply unsatisfying. What you can do, even a child can do; what you can't do, even Einstein couldn't have done. So Kaczynski's idea was to destroy existing institutions, get rid of all technology, and let people start over and work on hard problems anew.

Kaczynski's methods were crazy, but his loss of faith in the technological frontier is all around us. Consider the trivial but revealing hallmarks of urban hipsterdom: faux vintage photography, the handlebar mustache, and vinyl record players all hark back to an earlier time when people were still optimistic about the future. If everything worth doing has already been done, you may as well feign an allergy to achievement and become a barista.

Hipster or Unabomber?

All fundamentalists think this way, not just terrorists and hipsters. Religious fundamentalism, for example, allows no middle ground for hard questions: there are easy truths that children are expected to rattle off, and then there are the mysteries of God, which can't be explained. In between— the zone of hard truths—lies heresy. In the modern religion of environmentalism, the easy truth is that we must protect the environment. Beyond that, Mother Nature knows best, and she cannot be questioned. Free marketeers worship a similar logic. The value of things is set by the market. Even a child can look up stock quotes. But whether those prices make sense is not to be second-guessed; the market knows far more than you ever could.

Why has so much of our society come to believe that there are no hard secrets left? It might start with geography. There are no blank spaces left on the map anymore. If you grew up in the 18th century, there were still new places to go. After hearing tales of foreign adventure, you could become an explorer yourself. This was probably true up through the 19th and early 20th centuries; after that point photography from *National Geographic* showed every Westerner what even the most exotic, underexplored places on earth look like. Today, explorers are found mostly in history books and children's tales. Parents don't expect their kids to become explorers any more than they expect them to become pirates or sultans. Perhaps there are a few dozen uncontacted tribes somewhere deep in the Amazon, and we know there remains one last earthly frontier in the depths of the oceans. But the unknown seems less accessible than ever.

Along with the natural fact that physical frontiers have receded, four social trends have conspired to root out belief in secrets. First is incrementalism. From an early age, we are taught that the right way to do things is to proceed one very small step at a time, day by day, grade by grade. If you overachieve and end up learning something that's not on the test, you won't receive credit for it. But in exchange for doing exactly what's asked of you (and for doing it just a bit better than your peers), you'll get an A. This process extends all the way up through the tenure track, which is why academics usually chase large numbers of trivial publications instead of new frontiers.

Second is risk aversion. People are scared of secrets because they are scared of being wrong. By definition, a secret hasn't been vetted by the mainstream. If your goal is to never make a mistake in your life, you shouldn't look for secrets. The prospect of being lonely but right—dedicating your life to something that no one else believes in—is already hard. The prospect of being lonely and *wrong* can be unbearable.

Third is complacency. Social elites have the most freedom and ability to explore new thinking, but they seem to believe in secrets the least. Why search for a new secret if you can comfortably collect rents on everything that has already been done? Every fall, the deans at top law schools and business schools welcome the incoming class with the same implicit message: "You got into this elite institution. Your worries are over. You're set for life." But that's probably the kind of thing that's true only if you don't believe it.

Fourth is "flatness." As globalization advances, people perceive the world as one homogeneous, highly competitive marketplace: the world is "flat." Given that assumption, anyone who might have had the ambition to look for a secret will first ask himself: if it were possible to discover something new, wouldn't someone from the faceless global talent pool of smarter and more creative people have found it already? This voice of doubt can dissuade people from even starting to look for secrets in a world that seems too big a place for any individual to contribute something unique.

There's an optimistic way to describe the result of these

trends: today, you can't start a cult. Forty years ago, people were more open to the idea that not all knowledge was widely known. From the Communist Party to the Hare Krishnas, large numbers of people thought they could join some enlightened vanguard that would show them the Way. Very few people take unorthodox ideas seriously today, and the mainstream sees that as a sign of progress. We can be glad that there are fewer crazy cults now, yet that gain has come at great cost: we have given up our sense of wonder at secrets left to be discovered.

## THE WORLD ACCORDING TO CONVENTION

How must you see the world if you don't believe in secrets? You'd have to believe we've already solved all great questions. If today's conventions are correct, we can afford to be smug and complacent: "God's in His heaven, All's right with the world."

For example, a world without secrets would enjoy a perfect understanding of justice. Every injustice necessarily involves a moral truth that very few people recognize early on: in a democratic society, a wrongful practice persists only when most people don't perceive it to be unjust. At first, only a small minority of abolitionists knew that slavery was evil; that view has rightly become conventional, but it was still a secret in the early 19th century. To say that there are no secrets left today would mean that we live in a society with no hidden injustices.

In economics, disbelief in secrets leads to faith in efficient markets. But the existence of financial bubbles shows that markets can have extraordinary inefficiencies. (And the more people believe in efficiency, the bigger the bubbles get.) In 1999, nobody wanted to believe that the internet was irrationally overvalued. The same was true of housing in 2005: Fed chairman Alan Greenspan had to acknowledge some "signs of froth in local markets" but stated that "a bubble in home prices for the nation as a whole does not appear likely." The market reflected all knowable information and couldn't be questioned. Then home prices fell across the country, and the financial crisis of 2008 wiped out trillions. The future turned out to hold many secrets that economists could not make vanish simply by ignoring them.

What happens when a company stops believing in secrets? The sad decline of Hewlett-Packard provides a cautionary tale. In 1990, the company was worth $9 billion. Then came a decade of invention. In 1991, HP released the DeskJet 500C, the world's first affordable color printer. In 1993, it launched the OmniBook, one of the first "superportable" laptops. The next year, HP released the OfficeJet, the world's first all-in-one printer/fax/copier. This relentless product expansion paid off: by mid-2000, HP was worth $135 billion.

But starting in late 1999, when HP introduced a new branding campaign around the imperative to "invent," it stopped inventing things. In 2001, the company launched HP Services, a glorified consulting and support shop. In 2002, HP merged with Compaq, presumably because it didn't know what else to do. By 2005, the company's market cap had plunged to $70 billion—roughly half of what it had been just five years earlier.

HP's board was a microcosm of the dysfunction: it split into two factions, only one of which cared about new technology. That faction was led by Tom Perkins, an engineer who first came to HP in 1963 to run the company's research division at the personal request of Bill Hewlett and Dave Packard. At 73 years old in 2005, Perkins may as well have been a time-traveling visitor from a bygone age of optimism: he thought the board should identify the most promising new technologies and then have HP build them. But Perkins's faction lost out to its rival, led by chairwoman Patricia Dunn. A banker by trade, Dunn argued that charting a plan for future technology was beyond the board's competence. She thought the board should restrict itself to a night watchman's role: Was everything proper in the accounting department? Were people following all the rules?

Amid this infighting, someone on the board started leaking information to the press. When it was exposed that Dunn arranged a series of illegal wiretaps to identify the source, the backlash was worse than the original dissension, and the board was disgraced. Having abandoned the search for technological secrets, HP obsessed over gossip. As a result, by late 2012 HP was worth just $23 billion—not much more than it was worth in 1990, adjusting for inflation.

## THE CASE FOR SECRETS

You can't find secrets without looking for them. Andrew Wiles demonstrated this when he proved Fermat's Last Theorem after 358 years of fruitless inquiry by other mathematicians— the kind of sustained failure that might have suggested an inherently impossible task. Pierre de Fermat had conjectured in 1637 that no integers $a$, $b$, and $c$ could satisfy the equation $a^n + b^n = c^n$ for any integer $n$ greater than 2. He claimed to have a proof, but he died without writing it down, so his conjecture long remained a major unsolved problem in mathematics. Wiles started working on it in 1986, but he kept it a secret until 1993, when he knew he was nearing a solution. After nine years of hard work, Wiles proved the conjecture in 1995. He needed brilliance to succeed, but he also needed a faith in secrets. If you think something hard is impossible, you'll never even start trying to achieve it. Belief in secrets is an effective truth.

The actual truth is that there are many more secrets left to find, but they will yield only to relentless searchers. There is more to do in science, medicine, engineering, and in technology of all kinds. We are within reach not just of marginal goals set at the competitive edge of today's conventional disciplines, but of ambitions so great that even the boldest minds of the Scientific Revolution hesitated to announce them directly. We could cure cancer, dementia, and all the diseases of age and metabolic decay. We can find new ways to generate energy that free the world from conflict over fossil fuels. We can invent faster ways to travel from place to place over the surface of the planet; we can even learn how to escape it entirely and settle new frontiers. But we will never learn any of these secrets unless we demand to know them and force ourselves to look.

The same is true of business. Great companies can be built on open but unsuspected secrets about how the world works. Consider the Silicon Valley startups that have harnessed the spare capacity that is all around us but often ignored. Before Airbnb, travelers had little choice but to pay high prices for a hotel room, and property owners couldn't easily and reliably rent out their unoccupied space. Airbnb saw untapped supply and unaddressed demand where others saw nothing at all. The same is true of private car services Lyft and Uber. Few people imagined that it was possible to build a billion-dollar business by simply connecting people who want to go places with people willing to drive them there. We already had state-licensed taxicabs and private limousines; only by believing in and looking for secrets could you see beyond the convention to an opportunity hidden in plain sight. The same reason that so many internet companies, including Facebook, are often underestimated—their very simplicity—is itself an argument for secrets. If insights that look so elementary in retrospect can support important and valuable businesses, there must remain many great companies still to start.

## HOW TO FIND SECRETS

There are two kinds of secrets: secrets of nature and secrets about people. Natural secrets exist all around us; to find them, one must study some undiscovered aspect of the physical world. Secrets about people are different: they are things that people don't know about themselves or things they hide because they don't want others to know. So when thinking about what kind of company to build, there are two distinct questions to ask: What secrets is nature not telling you? What secrets are people not telling you?

It's easy to assume that natural secrets are the most

important: the people who look for them can sound intimidatingly authoritative. This is why physics PhDs are notoriously difficult to work with—because they know the most fundamental truths, they think they know *all* truths. But does understanding electromagnetic theory automatically make you a great marriage counselor? Does a gravity theorist know more about your business than you do? At PayPal, I once interviewed a physics PhD for an engineering job. Halfway through my first question, he shouted, "Stop! I already know what you're going to ask!" But he was wrong. It was the easiest no-hire decision I've ever made.

Secrets about people are relatively underappreciated. Maybe that's because you don't need a dozen years of higher education to ask the questions that uncover them: What are people not allowed to talk about? What is forbidden or taboo?

Sometimes looking for natural secrets and looking for human secrets lead to the same truth. Consider the monopoly secret again: *competition and capitalism are opposites.* If you didn't already know it, you could discover it the natural, empirical way: do a quantitative study of corporate profits and you'll see they're eliminated by competition. But you could also take the human approach and ask: what are people running companies not allowed to say? You would notice that monopolists downplay their monopoly status to avoid scrutiny, while competitive firms strategically exaggerate their uniqueness. The differences between firms only seem small on the surface; in fact, they are enormous.

The best place to look for secrets is where no one else is looking. Most people think only in terms of what they've been taught; schooling itself aims to impart conventional wisdom. So you might ask: are there any fields that matter but haven't been standardized and institutionalized? Physics, for example, is a real major at all major universities, and it's set in its ways. The opposite of physics might be astrology, but astrology doesn't matter. What about something like nutrition? Nutrition matters for everybody, but you can't major in it at Harvard. Most top scientists go into other fields. Most of the big studies were done 30 or 40 years ago, and most are seriously flawed. The food pyramid that told us to eat low fat and enormous amounts of grains was probably more a product of lobbying by Big Food than real science; its chief impact has been to aggravate our obesity epidemic. There's plenty more to learn: we know more about the physics of faraway stars than we know about human nutrition. It won't be easy, but it's not obviously impossible: exactly the kind of field that could yield secrets.

## WHAT TO DO WITH SECRETS

If you find a secret, you face a choice: Do you tell anyone? Or do you keep it to yourself?

It depends on the secret: some are more dangerous than others. As Faust tells Wagner:

> *The few who knew what might be learned,*
> *Foolish enough to put their whole heart on show,*
> *And reveal their feelings to the crowd below,*
> *Mankind has always crucified and burned.*

Unless you have perfectly conventional beliefs, it's rarely a good idea to tell everybody everything that you know.

So who do you tell? Whoever you need to, and no more. In practice, there's always a golden mean between telling nobody and telling everybody—and that's a company. The best entrepreneurs know this: every great business is built around a secret that's hidden from the outside. A great company is a conspiracy to change the world; when you share your secret, the recipient becomes a fellow conspirator.

As Tolkien wrote in *The Lord of the Rings*:

> *The Road goes ever on and on*
> *Down from the door where it began.*

Life is a long journey; the road marked out by the steps of previous travelers has no end in sight. But later on in the tale, another verse appears:

> *Still round the corner there may wait*
> *A new road or a secret gate,*
> *And though we pass them by today,*
> *Tomorrow we may come this way*
> *And take the hidden paths that run*
> *Towards the Moon or to the Sun.*

The road doesn't have to be infinite after all. Take the hidden paths.

# 9

# FOUNDATIONS

EVERY GREAT COMPANY is unique, but there are a few things that every business must get right at the beginning. I stress this so often that friends have teasingly nicknamed it "Thiel's law": *a startup messed up at its foundation cannot be fixed.*

Beginnings are special. They are qualitatively different from all that comes afterward. This was true 13.8 billion years ago, at the founding of our cosmos: in the earliest microseconds of its existence, the universe expanded by a factor of $10^{30}$—a million trillion trillion. As cosmogonic epochs came and went in those first few moments, the very laws of physics were different from those we know today.

It was also true 227 years ago at the founding of our country: fundamental questions were open for debate by the Framers during the few months they spent together at the Constitutional Convention. How much power should the central government have? How should representation in Congress be apportioned? Whatever your views on the compromises reached that summer in Philadelphia, they've been hard to change ever since: after ratifying the Bill of Rights in 1791, we've amended the Constitution only 17 times. Today, California has the same representation in the Senate as Alaska, even though it has more than 50 times as many people. Maybe that's a feature, not a bug. But we're probably stuck with it as long as the United States exists. Another constitutional convention is unlikely; today we debate only smaller questions.

Companies are like countries in this way. Bad decisions made early on—if you choose the wrong partners or hire the wrong people, for example—are very hard to correct after they are made. It may take a crisis on the order of bankruptcy before anybody will even try to correct them. As a founder, your first job is to get the first things right, because you cannot build a great company on a flawed foundation.

## FOUNDING MATRIMONY

When you start something, the first and most crucial decision you make is whom to start it with. Choosing a co-founder is like getting married, and founder conflict is just as ugly as divorce. Optimism abounds at the start of every relationship. It's unromantic to think soberly about what could go wrong, so people don't. But if the founders develop irreconcilable differences, the company becomes the victim.

In 1999, Luke Nosek was one of my co-founders at Pay-Pal, and I still work with him today at Founders Fund. But a year before PayPal, I invested in a company Luke started with someone else. It was his first startup; it was one of my first investments. Neither of us realized it then, but the venture was doomed to fail from the beginning because Luke and his co-founder were a terrible match. Luke is a brilliant and eccentric thinker; his co-founder was an MBA type who didn't want to miss out on the '90s gold rush. They met at a networking event, talked for a while, and decided to start a company together. That's no better than marrying the first person you meet at the slot machines in Vegas: you *might* hit the jackpot, but it probably won't work. Their company blew up and I lost my money.

Now when I consider investing in a startup, I study the founding teams. Technical abilities and complementary skill sets matter, but how well the founders know each other and how well they work together matter just as much. Founders should share a prehistory before they start a company together—otherwise they're just rolling dice.

## OWNERSHIP, POSSESSION, AND CONTROL

It's not just founders who need to get along. Everyone in your company needs to work well together. A Silicon Valley libertarian might say you could solve this problem by restricting yourself to a sole proprietorship. Freud, Jung, and every other psychologist has a theory about how every individual mind is divided against itself, but in business at least, working for yourself guarantees alignment. Unfortunately, it also limits what kind of company you can build. It's very hard to go from 0 to 1 without a team.

A Silicon Valley anarchist might say you could achieve perfect alignment as long as you hire just the right people, who will flourish peacefully without any guiding structure.

Serendipity and even free-form chaos at the workplace are supposed to help "disrupt" all the old rules made and obeyed by the rest of the world. And indeed, "if men were angels, no government would be necessary." But anarchic companies miss what James Madison saw: men aren't angels. That's why executives who manage companies and directors who govern them have separate roles to play; it's also why founders' and investors' claims on a company are formally defined. You need good people who get along, but you also need a structure to help keep everyone aligned for the long term.

To anticipate likely sources of misalignment in any company, it's useful to distinguish between three concepts:

- Ownership: who legally owns a company's equity?

- Possession: who actually runs the company on a day-to-day basis?

- Control: who formally governs the company's affairs?

A typical startup allocates ownership among founders, employees, and investors. The managers and employees who operate the company enjoy possession. And a board of directors, usually comprising founders and investors, exercises control.

In theory, this division works smoothly. Financial upside from part ownership attracts and rewards investors and workers. Effective possession motivates and empowers founders and employees—it means they can get stuff done. Oversight from the board places managers' plans in a broader perspective. In practice, distributing these functions among different people makes sense, but it also multiplies opportunities for misalignment.

To see misalignment at its most extreme, just visit the DMV. Suppose you need a new driver's license. Theoretically, it should be easy to get one. The DMV is a government agency, and we live in a democratic republic. All power resides in "the people," who elect representatives to serve them in government. If you're a citizen, you're a part owner of the DMV and your representatives control it, so you should be able to walk in and get what you need.

Of course, it doesn't work like that. We the people may "own" the DMV's resources, but that ownership is merely fictional. The clerks and petty tyrants who operate the DMV, however, enjoy very real possession of their small-time powers. Even the governor and the legislature charged with nominal control over the DMV can't change anything. The bureaucracy lurches ever sideways of its own inertia no matter what actions elected officials take. Accountable to no-body, the DMV is misaligned with everybody. Bureaucrats can make your licensing experience pleasurable or nightmarish at their sole discretion. You can try to bring up political theory and remind them that you are the boss, but that's unlikely to get you better service.

Big corporations do better than the DMV, but they're still prone to misalignment, especially between ownership and possession. The CEO of a huge company like General Motors, for example, will own some of the company's stock, but only a trivial portion of the total. Therefore he's incentivized to reward himself through the power of possession rather than the value of ownership. Posting good quarterly results will be enough for him to keep his high salary and corporate jet. Misalignment can creep in even if he receives stock compensation in the name of "shareholder value." If that stock comes as a reward for short-term performance, he will find it more lucrative and much easier to cut costs instead of investing in a plan that might create more value for all shareholders far in the future.

Unlike corporate giants, early-stage startups are small enough that founders usually have both ownership and possession. Most conflicts in a startup erupt between ownership and control—that is, between founders and investors on the board. The potential for conflict increases over time as interests diverge: a board member might want to take a company public as soon as possible to score a win for his venture firm, while the founders would prefer to stay private and grow the business.

In the boardroom, less is more. The smaller the board, the easier it is for the directors to communicate, to reach consensus, and to exercise effective oversight. However, that very effectiveness means that a small board can forcefully oppose management in any conflict. This is why it's crucial to choose wisely: every single member of your board matters. Even one problem director will cause you pain, and may even jeopardize your company's future.

A board of three is ideal. Your board should never exceed five people, unless your company is publicly held. (Government regulations effectively mandate that public companies have larger boards—the average is nine members.) By far the worst you can do is to make your board extra large. When unsavvy observers see a nonprofit organization with dozens of people on its board, they think: "Look how many great people are committed to this organization! It must be extremely well run." Actually, a huge board will exercise no effective oversight at all; it merely provides cover for whatever microdictator actually runs the organization. If you want that kind of free rein from your board, blow it up to giant size. If you want an effective board, keep it small.

## ON THE BUS OR OFF THE BUS

As a general rule, everyone you involve with your company should be involved full-time. Sometimes you'll have to break this rule; it usually makes sense to hire outside lawyers and accountants, for example. However, anyone who doesn't own stock options or draw a regular salary from your company is fundamentally misaligned. At the margin, they'll be biased to claim value in the near term, not help you create more in the future. That's why hiring consultants doesn't work. Part-time employees don't work. Even working remotely should be avoided, because misalignment can creep in whenever colleagues aren't together full-time, in the same place, every day. If you're deciding whether to bring someone on board, the decision is binary. Ken Kesey was right: you're either on the bus or off the bus.

## CASH IS NOT KING

For people to be fully committed, they should be properly compensated. Whenever an entrepreneur asks me to invest in his company, I ask him how much he intends to pay himself. A company does better the less it pays the CEO—that's one of the single clearest patterns I've noticed from investing in hundreds of startups. *In no case should a CEO of an early-stage, venture-backed startup receive more than $150,000 per year in salary.* It doesn't matter if he got used to making much more than that at Google or if he has a large mortgage and hefty private school tuition bills. If a CEO collects $300,000 per year, he risks becoming more like a politician than a founder. High pay incentivizes him to defend the status quo along with his salary, not to work with everyone else to surface problems and fix them aggressively. A cash-poor executive, by contrast, will focus on increasing the value of the company as a whole.

Low CEO pay also sets the standard for everyone else. Aaron Levie, the CEO of Box, was always careful to pay himself less than everyone else in the company—four years after he started Box, he was still living two blocks away from HQ in a one-bedroom apartment with no furniture except a mattress. Every employee noticed his obvious commitment to the company's mission and emulated it. If a CEO doesn't set an example by taking the *lowest* salary in the company, he can do the same thing by drawing the *highest* salary. So long as that figure is still modest, it sets an effective ceiling on cash compensation.

Cash is attractive. It offers pure optionality: once you get your paycheck, you can do anything you want with it. However, high cash compensation teaches workers to claim value from the company as it already exists instead of investing their time to create new value in the future. A cash bonus is slightly better than a cash salary—at least it's contingent on a job well done. But even so-called incentive pay encourages short-term thinking and value grabbing. Any kind of cash is more about the present than the future.

## VESTED INTERESTS

Startups don't need to pay high salaries because they can offer something better: part ownership of the company itself. Equity is the one form of compensation that can effectively orient people toward creating value in the future.

However, for equity to create commitment rather than conflict, you must allocate it very carefully. Giving everyone equal shares is usually a mistake: every individual has different talents and responsibilities as well as different opportunity costs, so equal amounts will seem arbitrary and unfair from the start. On the other hand, granting different amounts up front is just as sure to seem unfair. Resentment at this stage can kill a company, but there's no ownership formula to perfectly avoid it.

This problem becomes even more acute over time as more people join the company. Early employees usually get the most equity because they take more risk, but some later employees might be even more crucial to a venture's success. A secretary who joined eBay in 1996 might have made 200 times more than her industry-veteran boss who joined in 1999. The graffiti artist who painted Facebook's office walls in 2005 got stock that turned out to be worth $200 million, while a talented engineer who joined in 2010 might have made only $2 million. Since it's impossible to achieve perfect fairness when distributing ownership, founders would do well to keep the details secret. Sending out a company-wide email that lists everyone's ownership stake would be like dropping a nuclear bomb on your office.

Most people don't want equity at all. At PayPal, we once hired a consultant who promised to help us negotiate lucrative business development deals. The only thing he ever successfully negotiated was a $5,000 daily cash salary; he refused to accept stock options as payment. Stories of startup chefs becoming millionaires notwithstanding, people often find equity unattractive. It's not liquid like cash. It's tied to one specific company. And if that company doesn't succeed, it's worthless.

Equity is a powerful tool precisely because of these limitations. Anyone who prefers owning a part of your company to being paid in cash reveals a preference for the long term and a commitment to increasing your company's value in the future. Equity can't create perfect incentives, but it's the best way for a founder to keep everyone in the company broadly aligned.

## EXTENDING THE FOUNDING

Bob Dylan has said that he who is not busy being born is busy dying. If he's right, being born doesn't happen at just one moment—you might even continue to do it somehow, poetically at least. The founding moment of a company, however, really does happen just once: only at the very start do you have the opportunity to set the rules that will align people toward the creation of value in the future.

The most valuable kind of company maintains an openness

to invention that is most characteristic of beginnings. This leads to a second, less obvious understanding of the founding: it lasts as long as a company is creating new things, and it ends when creation stops. If you get the founding moment right, you can do more than create a valuable company: you can steer its distant future toward the creation of new things instead of the stewardship of inherited success. You might even extend its founding indefinitely.

# 10

## THE MECHANICS
## OF MAFIA

START WITH A THOUGHT EXPERIMENT: what would the ideal company culture look like? Employees should love their work. They should enjoy going to the office so much that formal business hours become obsolete and nobody watches the clock. The workspace should be open, not cubicled, and workers should feel at home: beanbag chairs and Ping-Pong tables might outnumber file cabinets. Free massages, on-site sushi chefs, and maybe even yoga classes would sweeten the scene. Pets should be welcome, too: perhaps employees' dogs and cats could come and join the office's tankful of tropical fish as unofficial company mascots.

What's wrong with this picture? It includes some of the absurd perks Silicon Valley has made famous, but none of the substance—and without substance perks don't work. You can't accomplish anything meaningful by hiring an interior decorator to beautify your office, a "human resources" consultant to fix your policies, or a branding specialist to hone your buzzwords. "Company culture" doesn't exist apart from the company itself: no company *has* a culture; every company *is* a culture. A startup is a team of people on a mission, and a good culture is just what that looks like on the inside.

## BEYOND PROFESSIONALISM

The first team that I built has become known in Silicon Valley as the "PayPal Mafia" because so many of my former colleagues have gone on to help each other start and invest in successful tech companies. We sold PayPal to eBay for $1.5 billion in 2002. Since then, Elon Musk has founded SpaceX and co-founded Tesla Motors; Reid Hoffman co-founded LinkedIn; Steve Chen, Chad Hurley, and Jawed Karim together founded YouTube; Jeremy Stoppelman and Russel Simmons founded Yelp; David Sacks co-founded Yammer; and I co-founded Palantir. Today all seven of those companies are worth more than $1 billion each. PayPal's office amenities never got much press, but the team has done extraordinarily well, both together and individually: the culture was strong enough to transcend the original company.

We didn't assemble a mafia by sorting through résumés and simply hiring the most talented people. I had seen the mixed results of that approach firsthand when I worked at a New York law firm. The lawyers I worked with ran a valuable business, and they were impressive individuals one by one. But the relationships between them were oddly thin. They spent all day together, but few of them seemed to have much to say to each other outside the office. Why work with a group of people who don't even like each other? Many seem to think it's a sacrifice necessary for making money. But taking a merely professional view of the workplace, in which free agents check in and out on a transactional basis, is worse than cold: it's not even rational. Since time is your most valuable asset, it's odd to spend it working with people who don't envision any long-term future together. If you can't count durable relationships among the fruits of your time at work, you haven't invested your time well—even in purely financial terms.

From the start, I wanted PayPal to be tightly knit instead of transactional. I thought stronger relationships would make us not just happier and better at work but also more successful in our careers even beyond PayPal. So we set out to hire people who would actually enjoy working together. They had to be talented, but even more than that they had to be excited about working specifically with us. That was the start of the PayPal Mafia.

## RECRUITING CONSPIRATORS

Recruiting is a core competency for any company. It should never be outsourced. You need people who are not just skilled on paper but who will work together cohesively after they're hired. The first four or five might be attracted by large equity stakes or high-profile responsibilities. More important than those obvious offerings is your answer to this question: *Why should the 20th employee join your company?*

Talented people don't *need* to work for you; they have plenty of options. You should ask yourself a more pointed version of the question: *Why would someone join your company as its 20th engineer when she could go work at Google for more money and more prestige?*

Here are some bad answers: "Your stock options will be worth more here than elsewhere." "You'll get to work with the smartest people in the world." "You can help solve the world's most challenging problems." What's wrong with valuable stock, smart people, or pressing problems? Nothing—but every company makes these same claims, so they won't help you stand out. General and undifferentiated pitches don't say ánything about why a recruit should join your company instead of many others.

The only good answers are specific to your company, so you won't find them in this book. But there are two general kinds of good answers: answers about your mission and answers about your team. You'll attract the employees you need if you can explain why your mission is compelling: not why it's important in general, but why you're doing something important that no one else is going to get done. That's the only thing that can make its importance unique. At PayPal, if you were excited by the idea of creating a new digital currency to replace the U.S. dollar, we wanted to talk to you; if not, you weren't the right fit.

However, even a great mission is not enough. The kind of recruit who would be most engaged as an employee will also wonder: "Are these the kind of people I want to work with?" You should be able to explain why your company is a unique match for him personally. And if you can't do that, he's probably not the right match.

Above all, don't fight the perk war. Anybody who would be more powerfully swayed by free laundry pickup or pet day care would be a bad addition to your team. Just cover the basics like health insurance and then promise what no others can: the opportunity to do irreplaceable work on a unique problem alongside great people. You probably can't be the Google of 2014 in terms of compensation or perks, but you *can* be like the Google of 1999 if you already have good answers about your mission and team.

## WHAT'S UNDER SILICON VALLEY'S HOODIES

*From the outside, everyone in your company should be different in the same way.*

Unlike people on the East Coast, who all wear the same skinny jeans or pinstripe suits depending on their industry, young people in Mountain View and Palo Alto go to work wearing T-shirts. It's a cliché that tech workers don't care about what they wear, but if you look closely at those T-shirts, you'll see the logos of the wearers' companies— and tech workers care about those very much. What makes a startup employee instantly distinguishable to outsiders is the branded T-shirt or hoodie that makes him look the same as his co-workers. The startup uniform encapsulates a simple but essential principle: everyone at your company should be different in the same way—a tribe of like-minded people fiercely devoted to the company's mission.

Max Levchin, my co-founder at PayPal, says that startups should make their early staff as personally similar as possible. Startups have limited resources and small teams. They must work quickly and efficiently in order to survive, and that's easier to do when everyone shares an understanding of the world. The early PayPal team worked well together because we were all the same kind of nerd. We all loved science fiction: *Cryptonomicon* was required reading, and we preferred the capitalist *Star Wars* to the communist *Star Trek*. Most important, we were all obsessed with creating a digital currency that would be controlled by individuals instead of governments. For the company to work, it didn't matter what people looked like or which country they came from, but we needed every new hire to be equally obsessed.

## DO ONE THING

*On the inside, every individual should be sharply distinguished by her work.*

When assigning responsibilities to employees in a startup, you could start by treating it as a simple optimization problem to efficiently match talents with tasks. But even if you could somehow get this perfectly right, any given solution would quickly break down. Partly that's because startups have to move fast, so individual roles can't remain static for long. But it's also because job assignments aren't just about the relationships between workers and tasks; they're also about relationships between employees.

The best thing I did as a manager at PayPal was to make every person in the company responsible for doing just one thing. Every employee's one thing was unique, and everyone knew I would evaluate him only on that one thing. I had started doing this just to simplify the task of managing people. But then I noticed a deeper result: defining roles reduced conflict. Most fights inside a company happen when colleagues compete for the same responsibilities. Startups face an especially high risk of this since job roles are fluid at the early stages. Eliminating competition makes it easier for everyone to build the kinds of long-term relationships that transcend mere professionalism. More than that, internal peace is what enables a startup to survive at all. When a startup fails, we often imagine it succumbing to predatory rivals in a competitive ecosystem. But every company is also its own ecosystem, and factional strife makes it vulnerable to outside threats. Internal conflict is like an autoimmune disease: the technical cause of death may be pneumonia, but the real cause remains hidden from plain view.

## OF CULTS AND CONSULTANTS

In the most intense kind of organization, members hang out only with other members. They ignore their families and abandon the outside world. In exchange, they experience strong feelings of belonging, and maybe get access to esoteric "truths" denied to ordinary people. We have a word for such organizations: cults. Cultures of total dedication look crazy from the outside, partly because the most notorious cults were homicidal: Jim Jones and Charles Manson did not make good exits.

But entrepreneurs should take cultures of extreme dedication seriously. Is a lukewarm attitude to one's work a sign of mental health? Is a merely professional attitude the only sane approach? The extreme opposite of a cult is a consulting firm like Accenture: not only does it lack a distinctive mission of its own, but individual consultants are regularly dropping in and out of companies to which they have no long-term connection whatsoever.

Every company culture can be plotted on a linear spectrum:

The best startups might be considered slightly less extreme kinds of cults. The biggest difference is that cults tend to be fanatically *wrong* about something important. People at a successful startup are fanatically *right* about something those outside it have missed. You're not going to learn those kinds of secrets from consultants, and you don't need to worry if your company doesn't make sense to conventional professionals. Better to be called a cult—or even a mafia.

# 11

## IF YOU BUILD IT, WILL THEY COME?

EVEN THOUGH SALES is everywhere, most people underrate its importance. Silicon Valley underrates it more than most. The geek classic *The Hitchhiker's Guide to the Galaxy* even explains the founding of our planet as a reaction against salesmen. When an imminent catastrophe requires the evacuation of humanity's original home, the population escapes on three giant ships. The thinkers, leaders, and achievers take the A Ship; the salespeople and consultants get the B Ship; and the workers and artisans take the C Ship. The B Ship leaves first, and all its passengers rejoice vainly. But the salespeople don't realize they are caught in a ruse: the A Ship and C Ship people had always thought that the B Ship people were useless, so they conspired to get rid of them. And it was the B Ship that landed on Earth.

Distribution may not matter in fictional worlds, but it matters in ours. We underestimate the importance of distribution—a catchall term for everything it takes to sell a product—because we share the same bias the A Ship and C Ship people had: salespeople and other "middlemen" supposedly get in the way, and distribution should flow magically from the creation of a good product. The *Field of Dreams* conceit is especially popular in Silicon Valley, where engineers are biased toward building cool stuff rather than selling it. But customers will not come just because you build it. You have to make that happen, and it's harder than it looks.

## NERDS VS. SALESMEN

The U.S. advertising industry collects annual revenues of $150 billion and employs more than 600,000 people. At $450 billion annually, the U.S. sales industry is even bigger. When they hear that 3.2 million Americans work in sales, seasoned executives will suspect the number is low, but engineers may sigh in bewilderment. What could that many salespeople possibly be doing?

In Silicon Valley, nerds are skeptical of advertising, marketing, and sales because they seem superficial and irrational. But advertising matters because it works. It works on nerds, and it works on *you*. You may think that you're an exception; that *your* preferences are authentic, and advertising only works on *other* people. It's easy to resist the most obvious sales pitches, so we entertain a false confidence in our own independence of mind. But advertising doesn't exist to make you buy a product right away; it exists to embed subtle impressions that will drive sales later. Anyone who can't acknowledge its likely effect on himself is doubly deceived.

Nerds are used to transparency. They add value by

becoming expert at a technical skill like computer programming. In engineering disciplines, a solution either works or it fails. You can evaluate someone else's work with relative ease, as surface appearances don't matter much. Sales is the opposite: an orchestrated campaign to change surface appearances without changing the underlying reality. This strikes engineers as trivial if not fundamentally dishonest. They know their own jobs are hard, so when they look at salespeople laughing on the phone with a customer or going to two-hour lunches, they suspect that no real work is being done. If anything, people overestimate the relative difficulty of science and engineering, because the challenges of those fields are obvious. What nerds miss is that it takes hard work to make sales look easy.

## SALES IS HIDDEN

All salesmen are actors: their priority is persuasion, not sincerity. That's why the word "salesman" can be a slur and the used car dealer is our archetype of shadiness. But we only react negatively to awkward, obvious salesmen—that is, the bad ones. There's a wide range of sales ability: there are many gradations between novices, experts, and masters. There are even sales grandmasters. If you don't know any grandmasters, it's not because you haven't encountered them, but rather because their art is hidden in plain sight. Tom Sawyer managed to persuade his neighborhood friends to whitewash the fence for him—a masterful move. But convincing them to actually *pay him* for the privilege of doing his chores was the move of a grandmaster, and his friends were none the wiser. Not much has changed since Twain wrote in 1876.

Like acting, sales works best when hidden. This explains why almost everyone whose job involves distribution— whether they're in sales, marketing, or advertising—has a job title that has nothing to do with those things. People who sell advertising are called "account executives." People who sell customers work in "business development." People who sell companies are "investment bankers." And people who sell themselves are called "politicians." There's a reason for these redescriptions: none of us wants to be reminded when we're being sold.

Whatever the career, sales ability distinguishes superstars from also-rans. On Wall Street, a new hire starts as an "analyst" wielding technical expertise, but his goal is to become a dealmaker. A lawyer prides himself on professional credentials, but law firms are led by the rainmakers who bring in big clients. Even university professors, who claim authority from scholarly achievement, are envious of the self-promoters who define their fields. Academic ideas about history or English don't just sell themselves on their intellectual merits. Even the agenda of fundamental physics and the future path of cancer research are results of persuasion. The most fundamental reason that even businesspeople underestimate the importance of sales is the systematic effort to hide it at every level of every field in a world secretly driven by it.

The engineer's grail is a product great enough that "it sells itself." But anyone who would actually say this about a real product must be lying: either he's delusional (lying to himself) or he's selling something (and thereby contradicting himself). The polar opposite business cliché warns that "the best product doesn't always win." Economists attribute this to "path dependence": specific historical circumstances independent of objective quality can determine which products enjoy widespread adoption. That's true, but it doesn't mean the operating systems we use today and the keyboard layouts on which we type were imposed by mere chance. It's better to think of distribution as something essential to the design of your product. If you've invented something new but you haven't invented an effective way to sell it, you have a bad business—no matter how good the product.

## HOW TO SELL A PRODUCT

Superior sales and distribution by itself can create a monopoly, even with no product differentiation. The converse is not true. No matter how strong your product—even if it easily fits into already established habits and anybody who tries it likes it immediately—you must still support it with a strong distribution plan.

Two metrics set the limits for effective distribution. The total net profit that you earn on average over the course of your relationship with a customer (Customer Lifetime Value, or CLV) must exceed the amount you spend on average to acquire a new customer (Customer Acquisition Cost, or CAC). In general, the higher the price of your product, the more you have to spend to make a sale—and the more it makes sense to spend it. Distribution methods can be plotted on a continuum:

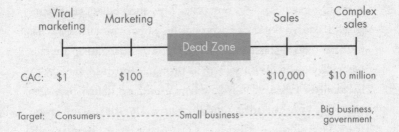

## Complex Sales

If your average sale is seven figures or more, every detail of every deal requires close personal attention. It might take months to develop the right relationships. You might make a sale only once every year or two. Then you'll usually have to follow up during installation and service the product long after the deal is done. It's hard to do, but this kind of "complex sales" is the only way to sell some of the most valuable products.

SpaceX shows that it can be done. Within just a few years of launching his rocket startup, Elon Musk persuaded NASA to sign billion-dollar contracts to replace the decommissioned space shuttle with a newly designed vessel from SpaceX. Politics matters in big deals just as much as technological ingenuity, so this wasn't easy. SpaceX employs more than 3,000 people, mostly in California. The traditional U.S. aerospace industry employs more than 500,000 people, spread throughout all 50 states. Unsurprisingly, members of Congress don't want to give up federal funds going to their home districts. But since complex sales requires making just a few deals each year, a sales grandmaster like Elon Musk can use that time to focus on the most crucial people—and even to overcome political inertia.

Complex sales works best when you don't have "salesmen" at all. Palantir, the data analytics company I co-founded with my law school classmate Alex Karp, doesn't employ anyone separately tasked with selling its product. Instead, Alex, who is Palantir's CEO, spends 25 days a month on the road, meeting with clients and potential clients. Our deal sizes range from $1 million to $100 million. At that price point, buyers want to talk to the CEO, not the VP of Sales.

Businesses with complex sales models succeed if they achieve 50% to 100% year-over-year growth over the course of a decade. This will seem slow to any entrepreneur dreaming of viral growth. You might expect revenue to increase 10x as soon as customers learn about an obviously superior product, but that almost never happens. Good enterprise sales strategy starts small, as it must: a new customer might agree to become your biggest customer, but they'll rarely be comfortable signing a deal completely out of scale with what you've sold before. Once you have a pool of reference customers who are successfully using your product, then you can begin the long and methodical work of hustling toward ever bigger deals.

## Personal Sales

Most sales are not particularly complex: average deal sizes might range between $10,000 and $100,000, and usually the CEO won't have to do all the selling himself. The challenge here isn't about how to make any particular sale, but how to establish a process by which a sales team of modest size can move the product to a wide audience.

In 2008, Box had a good way for companies to store their data safely and accessibly in the cloud. But people didn't know they needed such a thing—cloud computing hadn't caught on yet. That summer, Blake was hired as Box's third salesperson to help change that. Starting with small groups of users who had the most acute file sharing problems, Box's sales reps built relationships with more and more users in each client company. In 2009, Blake sold a small Box account to the Stanford Sleep Clinic, where researchers needed an easy, secure way to store experimental data logs. Today the university offers a Stanford-branded Box account to every one of its students and faculty members, and Stanford Hospital runs on Box. If it had started off by trying to sell the president of the university on an enterprise-wide solution, Box would have sold nothing. A complex sales approach would have made Box a forgotten startup failure; instead, personal sales made it a multibillion-dollar business.

Sometimes the product itself is a kind of distribution. Zoc-Doc is a Founders Fund portfolio company that helps people find and book medical appointments online. The company charges doctors a few hundred dollars per month to be included in its network. With an average deal size of just a few thousand dollars, ZocDoc needs lots of salespeople—so many that they have an internal recruiting team to do nothing but hire more. But making personal sales to doctors doesn't just bring in revenue; by adding doctors to the network, salespeople make the product more valuable to consumers (and more consumer users increases its appeal to doctors). More than 5 million people already use the service each month, and if it can continue to scale its network to include a majority of practitioners, it will become a fundamental utility for the U.S. health care industry.

## Distribution Doldrums

In between personal sales (salespeople obviously required) and traditional advertising (no salespeople required) there is a dead zone. Suppose you create a software service that helps convenience store owners track their inventory and manage ordering. For a product priced around $1,000, there might be no good distribution channel to reach the small businesses that might buy it. Even if you have a clear value proposition, how do you get people to hear it? Advertising would either be too broad (there's no TV channel that only convenience store owners watch) or too inefficient (on its own, an ad in *Convenience Store News* probably won't convince any owner to part with $1,000 a year). The product needs a personal sales effort, but at that price point, you simply don't have the resources to send an actual person to talk to every prospective customer. This is why so many small and medium-sized businesses don't use tools that bigger firms take for granted. It's not that small business proprietors are unusually backward or that good tools don't exist: distribution is the hidden bottleneck.

## Marketing and Advertising

Marketing and advertising work for relatively low-priced products that have mass appeal but lack any method of viral distribution. Procter & Gamble can't afford to pay salespeople to go door-to-door selling laundry detergent. (P&G

*does* employ salespeople to talk to grocery chains and large retail outlets, since one detergent sale made to these buyers might mean 100,000 one-gallon bottles.) To reach its end user, a packaged goods company has to produce television commercials, print coupons in newspapers, and design its product boxes to attract attention.

Advertising can work for startups, too, but only when your customer acquisition costs and customer lifetime value make every other distribution channel uneconomical. Consider e-commerce startup Warby Parker, which designs and sells fashionable prescription eyeglasses online instead of contracting sales out to retail eyewear distributors. Each pair starts at around $100, so assuming the average customer buys a few pairs in her lifetime, the company's CLV is a few hundred dollars. That's too little to justify personal attention on every transaction, but at the other extreme, hundred-dollar physical products don't exactly go viral. By running advertisements and creating quirky TV commercials, Warby is able to get its better, less expensive offerings in front of millions of eyeglass-wearing customers. The company states plainly on its website that "TV is a great big megaphone," and when you can only afford to spend dozens of dollars acquiring a new customer, you need the biggest megaphone you can find.

Every entrepreneur envies a recognizable ad campaign, but startups should resist the temptation to compete with bigger companies in the endless contest to put on the most memorable TV spots or the most elaborate PR stunts. I know this from experience. At PayPal we hired James Doohan, who played Scotty on *Star Trek,* to be our official spokesman.

When we released our first software for the PalmPilot, we invited journalists to an event where they could hear James recite this immortal line: "I've been beaming people up my whole career, but this is the first time I've ever been able to beam money!" It flopped—the few who actually came to cover the event weren't impressed. We were all nerds, so we had thought Scotty the Chief Engineer could speak with more authority than, say, Captain Kirk. (Just like a salesman, Kirk was always showboating out in some exotic locale and leaving it up to the engineers to bail him out of his own mistakes.) We were wrong: when Priceline.com cast William Shatner (the actor who played Kirk) in a famous series of TV spots, it worked for them. But by then Priceline was a major player. No early-stage startup can match big companies' advertising budgets. Captain Kirk truly is in a league of his own.

## Viral Marketing

A product is viral if its core functionality encourages users to invite their friends to become users too. This is how Facebook and PayPal both grew quickly: every time someone shares with a friend or makes a payment, they naturally invite more and more people into the network. This isn't just cheap—it's fast, too. If every new user leads to more than one additional user, you can achieve a chain reaction of exponential growth. The ideal viral loop should be as quick and frictionless as possible. Funny YouTube videos or internet memes get millions of views very quickly because they have extremely short cycle times: people see the kitten, feel warm inside, and forward it to their friends in a matter of seconds.

At PayPal, our initial user base was 24 people, all of whom worked at PayPal. Acquiring customers through banner advertising proved too expensive. However, by directly paying people to sign up and then paying them more to refer friends, we achieved extraordinary growth. This strategy cost us $20 per customer, but it also led to 7% daily growth, which meant that our user base nearly doubled every 10 days. After four or five months, we had hundreds of thousands of users and a viable opportunity to build a great company by servicing money transfers for small fees that ended up greatly exceeding our customer acquisition cost.

Whoever is first to dominate the most important segment of a market with viral potential will be the last mover in the whole market. At PayPal we didn't want to acquire more users at random; we wanted to get the most valuable users first. The most obvious market segment in email-based payments was the millions of emigrants still using Western Union to wire money to their families back home. Our product made that effortless, but the transactions were too infrequent. We needed a smaller niche market segment with a higher velocity of money—a segment we found in eBay "PowerSellers," the professional vendors who sold goods online through eBay's auction marketplace. There were 20,000 of them. Most had multiple auctions ending each day, and they bought almost as much as they sold, which meant a constant stream of payments. And because eBay's own solution to the payment problem was terrible, these merchants were extremely enthusiastic early adopters. Once PayPal dominated this segment and became *the* payments platform for eBay, there was no catching up—on eBay or anywhere else.

## The Power Law of Distribution

One of these methods is likely to be far more powerful than every other for any given business: distribution follows a power law of its own. This is counterintuitive for most entrepreneurs, who assume that more is more. But the kitchen sink approach—employ a few salespeople, place some magazine ads, and try to add some kind of viral functionality to the product as an afterthought—doesn't work. Most businesses get zero distribution channels to work: poor sales rather than bad product is the most common cause of failure. If you can get just one distribution channel to work, you have a great business. If you try for several but don't nail one, you're finished.

## Selling to Non-Customers

Your company needs to sell more than its product. You must also sell your company to employees and investors. There is a "human resources" version of the lie that great products sell themselves: "This company is so good that people will be clamoring to join it." And there's a fundraising version too: "This company is so great that investors will be banging down our door to invest." Clamor and frenzy are very real, but they rarely happen without calculated recruiting and pitching beneath the surface.

Selling your company to the media is a necessary part of selling it to everyone else. Nerds who instinctively mistrust the media often make the mistake of trying to ignore it. But just as you can never expect people to buy a superior product merely on its obvious merits without any distribution strategy, you should never assume that people will admire your company without a public relations strategy. Even if your particular product doesn't need media exposure to acquire customers because you have a viral distribution strategy, the press can help attract investors and employees. Any prospective employee worth hiring will do his own diligence; what he finds or doesn't find when he googles you will be critical to the success of your company.

## EVERYBODY SELLS

Nerds might wish that distribution could be ignored and salesmen banished to another planet. All of us want to believe that we make up our own minds, that sales doesn't work on us. But it's not true. Everybody has a product to sell—no matter whether you're an employee, a founder, or an investor. It's true even if your company consists of just you and your computer. Look around. If you don't see any salespeople, you're the salesperson.

# 12

# MAN AND MACHINE

A s mature industries stagnate, information technology has advanced so rapidly that it has now become synonymous with "technology" itself. Today, more than 1.5 billion people enjoy instant access to the world's knowledge using pocket-sized devices. Every one of today's smartphones has thousands of times more processing power than the computers that guided astronauts to the moon. And if Moore's law continues apace, tomorrow's computers will be even more powerful.

Computers already have enough power to outperform people in activities we used to think of as distinctively human. In 1997, IBM's Deep Blue defeated world chess champion Garry Kasparov. *Jeopardy!*'s best-ever contestant, Ken Jennings, succumbed to IBM's Watson in 2011. And Google's self-driving cars are already on California roads today. Dale Earnhardt Jr. needn't feel threatened by them, but the *Guardian* worries (on behalf of the millions of chauffeurs and cabbies in the world) that self-driving cars "could drive the next wave of unemployment."

Everyone expects computers to do more in the future—so much more that some wonder: 30 years from now, will there be anything left for people to do? "Software is eating the world," venture capitalist Marc Andreessen has announced with a tone of inevitability. VC Andy Kessler sounds almost gleeful when he explains that the best way to create productivity is "to get rid of people." *Forbes* captured a more anxious attitude when it asked readers: *Will a machine replace you?*

Futurists can seem like they hope the answer is yes. Luddites are so worried about being replaced that they would rather we stop building new technology altogether. Neither side questions the premise that better computers will necessarily replace human workers. But that premise is wrong: computers are complements for humans, not substitutes. The most valuable businesses of coming decades will be built by entrepreneurs who seek to empower people rather than try to make them obsolete.

## SUBSTITUTION VS. COMPLEMENTARITY

Fifteen years ago, American workers were worried about competition from cheaper Mexican substitutes. And that made sense, because humans really can substitute for each other. Today people think they can hear Ross Perot's "giant sucking sound" once more, but they trace it back to server farms somewhere in Texas instead of cut-rate factories in Tijuana. Americans fear technology in the near future because they see it as a replay of the globalization of the near past. But the situations are very different: people compete for jobs and for resources; computers compete for neither.

## Globalization Means Substitution

When Perot warned about foreign competition, both George H. W. Bush and Bill Clinton preached the gospel of free trade: since every person has a relative strength at some particular job, in theory the economy maximizes wealth when people specialize according to their advantages and then trade with each other. In practice, it's not unambiguously clear how well free trade has worked, for many workers at least. Gains from trade are greatest when there's a big discrepancy in comparative advantage, but the global supply of workers willing to do repetitive tasks for an extremely small wage is extremely large.

People don't just compete to supply labor; they also demand the same resources. While American consumers have benefited from access to cheap toys and textiles from China, they've had to pay higher prices for the gasoline newly desired by millions of Chinese motorists. Whether people eat shark fins in Shanghai or fish tacos in San Diego, they all need food and they all need shelter. And desire doesn't stop at subsistence—people will demand ever more as globalization continues. Now that millions of Chinese peasants can finally enjoy a secure supply of basic calories, they want more of them to come from pork instead of just grain. The convergence of desire is even more obvious at the top: all oligarchs have the same taste in Cristal, from Petersburg to Pyongyang.

## Technology Means Complementarity

Now think about the prospect of competition from computers instead of competition from human workers. On the supply side, computers are far more different from people than any two people are different from each other: men and machines are good at fundamentally different things. People have intentionality—we form plans and make decisions in complicated situations. We're less good at making sense of enormous amounts of data. Computers are exactly the opposite: they excel at efficient data processing, but they struggle to make basic judgments that would be simple for any human.

To understand the scale of this variance, consider another of Google's computer-for-human substitution projects. In 2012, one of their supercomputers made headlines when, after scanning 10 million thumbnails of YouTube videos, it learned to identify a cat with 75% accuracy. That seems impressive—until you remember that an average four-year-old can do it flawlessly. When a cheap laptop beats the

|  | SUPPLY<br>(of labor) | DEMAND<br>(for resources) |
|---|---|---|
| GLOBALIZATION<br>(other humans) | Substitution:<br>"The world<br>is flat." | Mimetic<br>consumer<br>competition |
| TECHNOLOGY<br>(better<br>computers) | Mostly<br>complementary | Machines<br>don't demand:<br>all value goes<br>to people |

smartest mathematicians at some tasks but even a supercomputer with 16,000 CPUs can't beat a child at others, you can tell that humans and computers are not just more or less powerful than each other—they're categorically different.

The stark differences between man and machine mean that gains from working with computers are much higher than gains from trade with other people. We don't trade with computers any more than we trade with livestock or lamps. And that's the point: computers are tools, not rivals.

The differences are even deeper on the demand side. Unlike people in industrializing countries, computers don't yearn for more luxurious foods or beachfront villas in Cap Ferrat; all they require is a nominal amount of electricity, which they're not even smart enough to want. When we design new computer technology to help solve problems, we get all the efficiency gains of a hyperspecialized trading partner without having to compete with it for resources. Properly understood, technology is the one way for us to escape competition in a globalizing world. As computers become more and more powerful, they won't be substitutes for humans: they'll be complements.

## COMPLEMENTARY BUSINESSES

Complementarity between computers and humans isn't just a macro-scale fact. It's also the path to building a great business. I came to understand this from my experience at PayPal. In mid-2000, we had survived the dot-com crash and we were growing fast, but we faced one huge problem: we were losing upwards of $10 million to credit card fraud every month. Since we were processing hundreds or even thousands of transactions per minute, we couldn't possibly review each one—no human quality control team could work that fast.

So we did what any group of engineers would do: we tried to automate a solution. First, Max Levchin assembled an elite team of mathematicians to study the fraudulent transfers in detail. Then we took what we learned and wrote software to automatically identify and cancel bogus transactions in real time. But it quickly became clear that this approach wouldn't work either: after an hour or two, the thieves would catch on and change their tactics. We were dealing with an adaptive enemy, and our software couldn't adapt in response.

The fraudsters' adaptive evasions fooled our automatic detection algorithms, but we found that they didn't fool our human analysts as easily. So Max and his engineers rewrote the software to take a hybrid approach: the computer would flag the most suspicious transactions on a well-designed user interface, and human operators would make the final judgment as to their legitimacy. Thanks to this hybrid system—we named it "Igor," after the Russian fraudster who bragged that we'd never be able to stop him—we turned our first quarterly profit in the first quarter of 2002 (as opposed to a quarterly loss of $29.3 million one year before). The FBI asked us if we'd let them use Igor to help detect financial crime. And Max was able to boast, grandiosely but truthfully, that he was "the Sherlock Holmes of the Internet Underground."

This kind of man-machine symbiosis enabled PayPal to stay in business, which in turn enabled hundreds of thousands of small businesses to accept the payments they needed to thrive on the internet. None of it would have been possible without the man-machine solution—even though most people would never see it or even hear about it.

I continued to think about this after we sold PayPal in 2002: if humans and computers together could achieve dramatically better results than either could attain alone, what other valuable businesses could be built on this core principle? The next year, I pitched Alex Karp, an old Stanford classmate, and Stephen Cohen, a software engineer, on a new startup idea: we would use the human-computer hybrid approach from PayPal's security system to identify terrorist networks and financial fraud. We already knew the FBI was interested, and in 2004 we founded Palantir, a software company that helps people extract insight from divergent sources of information. The company is on track to book sales of $1 billion in 2014, and *Forbes* has called Palantir's software the "killer app" for its rumored role in helping the government locate Osama bin Laden.

We have no details to share from that operation, but we can say that neither human intelligence by itself nor computers alone will be able to make us safe. America's two biggest spy agencies take opposite approaches: The Central Intelligence Agency is run by spies who privilege humans. The National Security Agency is run by generals who prioritize computers. CIA analysts have to wade through so much noise that it's very difficult to identify the most serious threats. NSA computers can process huge quantities of data, but machines alone cannot authoritatively determine whether someone is plotting a terrorist act. Palantir aims to transcend these opposing biases: its software analyzes the data the government feeds it—phone records of radical clerics in Yemen or bank accounts linked to terror cell activity, for instance—and flags suspicious activities for a trained analyst to review.

In addition to helping find terrorists, analysts using Palantir's software have been able to predict where insurgents plant IEDs in Afghanistan; prosecute high-profile insider trading cases; take down the largest child pornography ring in the world; support the Centers for Disease Control and Prevention in fighting foodborne disease outbreaks; and save both commercial banks and the government hundreds of millions of dollars annually through advanced fraud detection.

Advanced software made this possible, but even more important were the human analysts, prosecutors, scientists, and financial professionals without whose active engagement the software would have been useless.

Think of what professionals do in their jobs today. Lawyers must be able to articulate solutions to thorny problems in several different ways—the pitch changes depending on whether you're talking to a client, opposing counsel, or a judge. Doctors need to marry clinical understanding with an ability to communicate it to non-expert patients. And good teachers aren't just experts in their disciplines: they must also understand how to tailor their instruction to different individuals' interests and learning styles. Computers might be able to do some of these tasks, but they can't combine them effectively. Better technology in law, medicine, and education won't replace professionals; it will allow them to do even more.

LinkedIn has done exactly this for recruiters. When LinkedIn was founded in 2003, they didn't poll recruiters to find discrete pain points in need of relief. And they didn't try to write software that would replace recruiters outright. Recruiting is part detective work and part sales: you have to scrutinize applicants' history, assess their motives and compatibility, and persuade the most promising ones to join you. Effectively replacing all those functions with a computer would be impossible. Instead, LinkedIn set out to transform how recruiters did their jobs. Today, more than 97% of recruiters use LinkedIn and its powerful search and filtering functionality to source job candidates, and the network also creates value for the hundreds of millions of professionals who use it to manage their personal brands. If LinkedIn had tried to simply replace recruiters with technology, they wouldn't have a business today.

## The Ideology of Computer Science

Why do so many people miss the power of complementarity? It starts in school. Software engineers tend to work on projects that replace human efforts because that's what they're trained to do. Academics make their reputations through specialized research; their primary goal is to publish papers, and publication means respecting the limits of a particular discipline. For computer scientists, that means reducing human capabilities into specialized tasks that computers can be trained to conquer one by one.

Just look at the trendiest fields in computer science today. The very term "machine learning" evokes imagery of replacement, and its boosters seem to believe that computers can be taught to perform almost any task, so long as we feed them enough training data. Any user of Netflix or Amazon has experienced the results of machine learning firsthand: both companies use algorithms to recommend products based on your viewing and purchase history. Feed them more data and the recommendations get ever better. Google Translate works the same way, providing rough but serviceable translations into any of the 80 languages it supports—not because the software understands human language, but because it has extracted patterns through statistical analysis of a huge corpus of text.

The other buzzword that epitomizes a bias toward substitution is "big data." Today's companies have an insatiable appetite for data, mistakenly believing that more data always creates more value. But big data is usually dumb data. Computers can find patterns that elude humans, but they don't know how to compare patterns from different sources or how to interpret complex behaviors. Actionable insights can only come from a human analyst (or the kind of generalized artificial intelligence that exists only in science fiction).

We have let ourselves become enchanted by big data only because we exoticize technology. We're impressed with small feats accomplished by computers alone, but we ignore big achievements from complementarity because the human contribution makes them less uncanny. Watson, Deep Blue, and ever-better machine learning algorithms are cool. But the most valuable companies in the future won't ask what problems can be solved with computers alone. Instead, they'll ask: *how can computers help humans solve hard problems?*

## EVER-SMARTER COMPUTERS: FRIEND OR FOE?

The future of computing is necessarily full of unknowns. It's become conventional to see ever-smarter anthropomorphized robot intelligences like Siri and Watson as harbingers of things to come; once computers can answer all our questions, perhaps they'll ask why they should remain subservient to us at all.

The logical endpoint to this substitutionist thinking is called "strong AI": computers that eclipse humans on every important dimension. Of course, the Luddites are terrified by the possibility. It even makes the futurists a little uneasy; it's not clear whether strong AI would save humanity or doom it. Technology is supposed to *increase* our mastery over nature and *reduce* the role of chance in our lives; building smarter-than-human computers could actually bring chance back with a vengeance. Strong AI is like a cosmic lottery ticket: if we win, we get utopia; if we lose, Skynet substitutes us out of existence.

But even if strong AI is a real possibility rather than an imponderable mystery, it won't happen anytime soon: replacement by computers is a worry for the 22nd century. Indefinite fears about the far future shouldn't stop us from making definite plans today. Luddites claim that we shouldn't build the computers that might replace people someday; crazed futurists argue that we should. These two positions are mutually

# THE FUTURE OF STRONG AI?

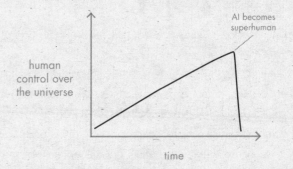

exclusive but they are not exhaustive: there is room in between for sane people to build a vastly better world in the decades ahead. As we find new ways to use computers, they won't just get better at the kinds of things people already do; they'll help us to do what was previously unimaginable.

# 13

## SEEING GREEN

A T THE START of the 21st century, everyone agreed that the next big thing was clean technology. It had to be: in Beijing, the smog had gotten so bad that people couldn't see from building to building—even breathing was a health risk. Bangladesh, with its arsenic-laden water wells, was suffering what the *New York Times* called "the biggest mass poisoning in history." In the U.S., Hurricanes Ivan and Katrina were said to be harbingers of the coming devastation from global warming. Al Gore implored us to attack these problems "with the urgency and resolve that has previously been seen only when nations mobilized for war." People got busy: entrepreneurs started thousands of cleantech companies, and investors poured more than $50 billion into them. So began the quest to cleanse the world.

It didn't work. Instead of a healthier planet, we got a massive cleantech bubble. Solyndra is the most famous green ghost, but most cleantech companies met similarly disastrous ends—more than 40 solar manufacturers went out of business or filed for bankruptcy in 2012 alone. The leading index of alternative energy companies shows the bubble's dramatic deflation:

## RENIXX (RENEWABLE ENERGY INDUSTRIAL INDEX)

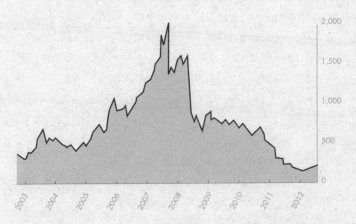

Why did cleantech fail? Conservatives think they already know the answer: as soon as green energy became a priority for the government, it was poisoned. But there really were (and there still are) good reasons for making energy a priority. And the truth about cleantech is more complex and more important than government failure. Most cleantech companies crashed because they neglected one or more of the seven questions that every business must answer:

1. The Engineering Question

   *Can you create breakthrough technology instead of incremental improvements?*

## 2. The Timing Question

*Is now the right time to start your particular business?*

## 3. The Monopoly Question

*Are you starting with a big share of a small market?*

## 4. The People Question

*Do you have the right team?*

## 5. The Distribution Question

*Do you have a way to not just create but deliver your product?*

## 6. The Durability Question

*Will your market position be defensible 10 and 20 years into the future?*

## 7. The Secret Question

*Have you identified a unique opportunity that others don't see?*

We've discussed these elements before. Whatever your industry, any great business plan must address every one of them. If you don't have good answers to these questions, you'll run into lots of "bad luck" and your business will fail. If you nail all seven, you'll master fortune and succeed. Even getting five or six correct might work. But the striking thing about the cleantech bubble was that people were starting companies with zero good answers—and that meant hoping for a miracle.

It's hard to know exactly why any particular cleantech company failed, since almost all of them made several serious mistakes. But since *any one* of those mistakes is enough to doom your company, it's worth reviewing cleantech's losing scorecard in more detail.

## THE ENGINEERING QUESTION

A great technology company should have proprietary technology an order of magnitude better than its nearest substitute. But cleantech companies rarely produced 2x, let alone 10x, improvements. Sometimes their offerings were actually *worse* than the products they sought to replace. Solyndra developed novel, cylindrical solar cells, but to a first approximation, cylindrical cells are only $1/\pi$ as efficient as flat ones—they simply don't receive as much direct sunlight. The company tried to correct for this deficiency by using mirrors to reflect more sunlight to hit the bottoms of the panels, but it's hard to recover from a radically inferior starting point.

Companies must strive for 10x better because merely incremental improvements often end up meaning no improvement at all for the end user. Suppose you develop a new wind turbine that's 20% more efficient than any existing technology—when you test it in the laboratory. That sounds good at first, but the lab result won't begin to compensate for the expenses and risks faced by any new product in the real world. And even if your system really is 20% better on net for the customer who buys it, people are so used to exaggerated claims that you'll be met with skepticism when you try to sell it. Only when your product is 10x better can you offer the customer transparent superiority.

## THE TIMING QUESTION

Cleantech entrepreneurs worked hard to convince themselves that their appointed hour had arrived. When he announced his new company in 2008, SpectraWatt CEO Andrew Wilson stated that "[t]he solar industry is akin to where the microprocessor industry was in the late 1970s. There is a lot to be figured out and improved." The second part was right, but the microprocessor analogy was way off. Ever since the first microprocessor was built in 1970, computing advanced not just rapidly but exponentially. Look at Intel's early product release history:

| Generation | Processor Model | Year |
|---|---|---|
| 4-bit | 4004 | 1971 |
| 8-bit | 8008 | 1972 |
| 16-bit | 8086 | 1978 |
| 32-bit | iAPX 432 | 1981 |

The first silicon solar cell, by contrast, was created by Bell Labs in 1954—more than *a half century* before Wilson's press release. Photovoltaic efficiency improved in the intervening decades, but slowly and linearly: Bell's first solar cell had about 6% efficiency; neither today's crystalline silicon cells nor modern thin-film cells have exceeded 25% efficiency in the field. There were few engineering developments in the mid-2000s to suggest impending liftoff. Entering a slow-moving market can be a good strategy, but only if you have a definite and realistic plan to take it over. The failed cleantech companies had none.

## THE MONOPOLY QUESTION

In 2006, billionaire technology investor John Doerr announced that "green is the new red, white and blue." He could have stopped at "red." As Doerr himself said, "Internet-sized markets are in the billions of dollars; the energy markets are in the trillions." What he didn't say is that huge, trillion-dollar markets mean ruthless, bloody competition. Others echoed Doerr over and over: in the 2000s, I listened to dozens of cleantech entrepreneurs begin fantastically rosy PowerPoint presentations with all-too-true tales of trillion-dollar markets—as if that were a good thing.

Cleantech executives emphasized the bounty of an energy market big enough for all comers, but each one typically believed that *his own* company had an edge. In 2006, Dave Pearce, CEO of solar manufacturer MiaSolé, admitted to a congressional panel that his company was just one of several "very strong" startups working on one particular kind of thin-film solar cell development. Minutes later, Pearce predicted that MiaSolé would become "the largest producer of thin-film solar cells in the world" within a year's time. That didn't happen, but it might not have helped them anyway: thin-film is just one of more than a dozen kinds of solar cells. Customers won't care about any particular technology unless it solves a particular problem in a superior way. And if you can't monopolize a unique solution for a small market, you'll be stuck with vicious competition. That's what happened to MiaSolé, which was acquired in 2013 for hundreds of millions of dollars less than its investors had put into the company.

Exaggerating your own uniqueness is an easy way to botch the monopoly question. Suppose you're running a solar company that's successfully installed hundreds of solar panel systems with a combined power generation capacity of 100 megawatts. Since total U.S. solar energy production capacity is 950 megawatts, you own 10.53% of the market. Congratulations, you tell yourself: you're a player.

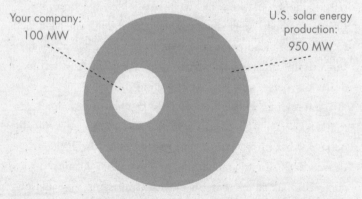

Your company:
100 MW

U.S. solar energy
production:
950 MW

But what if the U.S. solar energy market isn't the relevant market? What if the relevant market is the *global* solar market, with a production capacity of 18 gigawatts? Your 100 megawatts now makes you a very small fish indeed: suddenly you own less than 1% of the market.

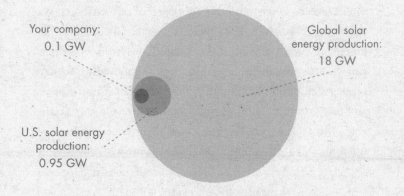

Your company:
0.1 GW

Global solar
energy production:
18 GW

U.S. solar energy
production:
0.95 GW

And what if the appropriate measure isn't global solar, but rather renewable energy *in general*? Annual production capacity from renewables is 420 gigawatts globally; you just shrank to 0.02% of the market. And compared to the total global power generation capacity of 15,000 gigawatts, your 100 megawatts is just a drop in the ocean.

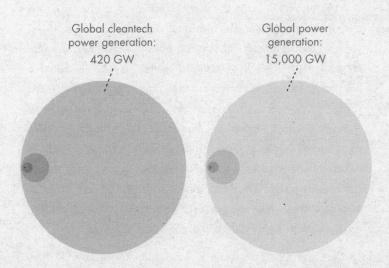

Global cleantech
power generation:
420 GW

Global power
generation:
15,000 GW

Cleantech entrepreneurs' thinking about markets was hopelessly confused. They would rhetorically shrink their market in order to seem differentiated, only to turn around and ask to be valued based on huge, supposedly lucrative markets. But you can't dominate a submarket if it's fictional, and huge markets are highly competitive, not highly attainable. Most cleantech founders would have been better off opening a new British restaurant in downtown Palo Alto.

## THE PEOPLE QUESTION

Energy problems are engineering problems, so you would expect to find nerds running cleantech companies. You'd be wrong: the ones that failed were run by shockingly nontechnical teams. These salesman-executives were good at raising capital and securing government subsidies, but they were less good at building products that customers wanted to buy.

At Founders Fund, we saw this coming. The most obvious clue was sartorial: cleantech executives were running around wearing suits and ties. This was a huge red flag, because real technologists wear T-shirts and jeans. So we instituted a blanket rule: pass on any company whose founders dressed up for pitch meetings. Maybe we still would have avoided these bad investments if we had taken the time to evaluate each company's technology in detail. But the team insight— never invest in a tech CEO that wears a suit—got us to the truth a lot faster. The best sales is hidden. There's nothing wrong with a CEO who can sell, but if he actually *looks* like a salesman, he's probably bad at sales and worse at tech.

Solyndra CEO Brian Harrison; Tesla Motors CEO Elon Musk

## THE DISTRIBUTION QUESTION

Cleantech companies effectively courted government and investors, but they often forgot about customers. They learned
the hard way that the world is not a laboratory: selling and
delivering a product is at least as important as the product
itself.

Just ask Israeli electric vehicle startup Better Place, which
from 2007 to 2012 raised and spent more than $800 million
to build swappable battery packs and charging stations for
electric cars. The company sought to "create a green alternative that would lessen our dependence on highly polluting
transportation technologies." And it did just that—at least by
1,000 cars, the number it sold before filing for bankruptcy.
Even selling that many was an achievement, because each of
those cars was very hard for customers to buy.

For starters, it was never clear what you were actually buying. Better Place bought sedans from Renault and refitted them with electric batteries and electric motors. So, were you buying an electric Renault, or were you buying a Better Place? In any case, if you decided to buy one, you had to jump through a series of hoops. First, you needed to seek approval from Better Place. To get that, you had to prove that you lived close enough to a Better Place battery swapping station and promise to follow predictable routes. If you passed that test, you had to sign up for a fueling subscription in order to recharge your car. Only then could you get started learning the new behavior of stopping to swap out battery packs on the road.

Better Place thought its technology spoke for itself, so they didn't bother to market it clearly. Reflecting on the company's failure, one frustrated customer asked, "Why wasn't there a billboard in Tel Aviv showing a picture of a Toyota Prius for 160,000 shekels and a picture of this car, for 160,000 plus fuel for four years?" *He* still bought one of the cars, but unlike most people, he was a hobbyist who "would do anything to keep driving it." Unfortunately, he can't: as the Better Place board of directors stated upon selling the company's assets for a meager $12 million in 2013, "The technical challenges we overcame successfully, but the other obstacles we were not able to overcome."

## THE DURABILITY QUESTION

Every entrepreneur should plan to be the last mover in her particular market. That starts with asking yourself: what will the world look like 10 and 20 years from now, and how will my business fit in?

Few cleantech companies had a good answer. As a result, all their obituaries resemble each other. A few months before it filed for bankruptcy in 2011, Evergreen Solar explained its decision to close one of its U.S. factories:

> Solar manufacturers in China have received considerable government and financial support. . . . Although [our] production costs . . . are now below originally planned levels and lower than most western manufacturers, they are still much higher than those of our low cost competitors in China.

But it wasn't until 2012 that the "blame China" chorus really exploded. Discussing its bankruptcy filing, U.S. Department of Energy–backed Abound Solar blamed "aggressive pricing actions from Chinese solar panel companies" that "made it very difficult for an early stage startup company . . . to scale in current market conditions." When solar panel maker Energy Conversion Devices failed in February 2012, it went beyond blaming China in a press release and filed a $950 million lawsuit against three prominent Chinese solar manufacturers—the same companies that Solyndra's trustees in bankruptcy sued later that year on the grounds of attempted monopolization, conspiracy, and predatory pricing. But was competition from Chinese manufacturers really impossible to predict? Cleantech entrepreneurs would have done well to rephrase the durability question and ask: what will stop China from wiping out my business? Without an answer, the result shouldn't have come as a surprise.

Beyond the failure to anticipate competition in manufacturing the same green products, cleantech embraced misguided assumptions about the energy market as a whole. An industry premised on the supposed twilight of fossil fuels was blindsided by the rise of fracking. In 2000, just 1.7% of America's natural gas came from fracked shale. Five years later, that figure had climbed to 4.1%. Nevertheless, nobody in cleantech took this trend seriously: renewables were the only way forward; fossil fuels couldn't *possibly* get cheaper or cleaner in the future. But they did. By 2013, shale gas accounted for 34% of America's natural gas, and gas prices had fallen more than 70% since 2008, devastating most renewable energy business models. Fracking may not be a durable energy solution, either, but it was enough to doom cleantech companies that didn't see it coming.

## THE SECRET QUESTION

Every cleantech company justified itself with conventional truths about the need for a cleaner world. They deluded themselves into believing that an overwhelming social need for alternative energy solutions implied an overwhelming business opportunity for cleantech companies of all kinds. Consider how conventional it had become by 2006 to be bullish on solar. That year, President George W. Bush heralded a future of "solar roofs that will enable the American family to be able to generate their own electricity." Investor and cleantech executive Bill Gross declared that the "potential for solar is enormous." Suvi Sharma, then-CEO of solar manufacturer Solaria, admitted that while "there is a gold rush feeling" to solar, "there's also real gold here—or, in our case, sunshine." But rushing to embrace the convention sent scores of solar panel companies—Q-Cells, Evergreen Solar, SpectraWatt, and even Gross's own Energy Innovations, to name just a few—from promising beginnings to bankruptcy court very quickly. Each of the casualties had described their bright futures using broad conventions on which everybody agreed. Great companies have secrets: specific reasons for success that other people don't see.

## THE MYTH OF SOCIAL ENTREPRENEURSHIP

Cleantech entrepreneurs aimed for more than just success as most businesses define it. The cleantech bubble was the biggest phenomenon—and the biggest flop—in the history of "social entrepreneurship." This philanthropic approach to business starts with the idea that corporations and nonprofits have until now been polar opposites: corporations have great power, but they're shackled to the profit motive; nonprofits pursue the public interest, but they're weak players in the wider economy. Social entrepreneurs aim to combine the best of both worlds and "do well by doing good." Usually they end up doing neither.

The ambiguity between social and financial goals doesn't help. But the ambiguity in the word "social" is even more of a problem: if something is "socially good," is it good *for* society, or merely *seen* as good *by* society? Whatever is good enough to receive applause from all audiences can only be conventional, like the general idea of green energy.

Progress isn't held back by some difference between

corporate greed and nonprofit goodness; instead, we're held back by the sameness of both. Just as corporations tend to copy each other, nonprofits all tend to push the same priorities. Cleantech shows the result: hundreds of undifferentiated products all in the name of one overbroad goal.

Doing something *different* is what's truly good for society— and it's also what allows a business to profit by monopolizing a new market. The best projects are likely to be overlooked, not trumpeted by a crowd; the best problems to work on are often the ones nobody else even tries to solve.

## TESLA: 7 FOR 7

Tesla is one of the few cleantech companies started last decade to be thriving today. They rode the social buzz of cleantech better than anyone, but they got the seven questions right, so their success is instructive:

> TECHNOLOGY. Tesla's technology is so good that other car companies rely on it: Daimler uses Tesla's battery packs; Mercedes-Benz uses a Tesla powertrain; Toyota uses a Tesla motor. General Motors has even created a task force to track Tesla's next moves. But Tesla's greatest technological achievement isn't any single part or component, but rather its ability to integrate many components into one superior product. The Tesla Model S sedan, elegantly designed from end to end, is more than the sum of its parts: *Consumer Reports* rated it higher than any other car

ever reviewed, and both *Motor Trend* and *Automobile* magazines named it their 2013 Car of the Year.

TIMING. In 2009, it was easy to think that the government would continue to support cleantech: "green jobs" were a political priority, federal funds were already earmarked, and Congress even seemed likely to pass cap-and-trade legislation. But where others saw generous subsidies that could flow indefinitely, Tesla CEO Elon Musk rightly saw a one-time-only opportunity. In January 2010—about a year and a half before Solyndra imploded under the Obama administration and politicized the subsidy question—Tesla secured a $465 million loan from the U.S. Department of Energy. A half-billion-dollar subsidy was unthinkable in the mid-2000s. It's unthinkable today. There was only one moment where that was possible, and Tesla played it perfectly.

MONOPOLY. Tesla started with a tiny submarket that it could dominate: the market for high-end electric sports cars. Since the first Roadster rolled off the production line in 2008, Tesla's sold only about 3,000 of them, but at $109,000 apiece that's not trivial. Starting small allowed Tesla to undertake the necessary R&D to build the slightly less expensive Model S, and now Tesla owns the luxury electric sedan market, too. They sold more than 20,000 sedans in 2013 and now Tesla is in prime position to expand to broader markets in the future.

TEAM. Tesla's CEO is the consummate engineer *and* salesman, so it's not surprising that he's assembled a team that's very good at both. Elon describes his staff this way: "If you're at Tesla, you're choosing to be at the equivalent of Special Forces. There's the regular army, and that's fine, but if you are working at Tesla, you're choosing to step up your game."

DISTRIBUTION. Most companies underestimate distribution, but Tesla took it so seriously that it decided to own the entire distribution chain. Other car companies are beholden to independent dealerships: Ford and Hyundai make cars, but they rely on other people to sell them. Tesla sells and services its vehicles in its own stores. The up-front costs of Tesla's approach are much higher than traditional dealership distribution, but it affords control over the customer experience, strengthens Tesla's brand, and saves the company money in the long run.

DURABILITY. Tesla has a head start and it's moving faster than anyone else—and that combination means its lead is set to widen in the years ahead. A coveted brand is the clearest sign of Tesla's breakthrough: a car is one of the biggest purchasing decisions that people ever make, and consumers' trust in that category is hard to win. And unlike every other car company, at Tesla the founder is still in charge, so it's not going to ease off anytime soon.

SECRETS. Tesla knew that fashion drove interest in cleantech. Rich people especially wanted to appear "green," even if it meant driving a boxy Prius or clunky Honda Insight. Those cars only made drivers look cool by association with the famous eco-conscious movie stars who owned them as well. So Tesla decided to build cars that made drivers look cool, period—Leonardo DiCaprio even ditched his Prius for an expensive (and expensive-looking) Tesla Roadster. While generic cleantech companies struggled to differentiate themselves, Tesla built a unique brand around the secret that cleantech was even more of a social phenomenon than an environmental imperative.

## ENERGY 2.0

Tesla's success proves that there was nothing inherently wrong with cleantech. The biggest idea behind it is right: the world really will need new sources of energy. Energy is the master resource: it's how we feed ourselves, build shelter, and make everything we need to live comfortably. Most of the world dreams of living as comfortably as Americans do today, and globalization will cause increasingly severe energy challenges unless we build new technology. There simply aren't enough resources in the world to replicate old approaches or redistribute our way to prosperity.

Cleantech gave people a way to be optimistic about the future of energy. But when indefinitely optimistic investors betting on the general idea of green energy funded cleantech companies that lacked specific business plans, the result was a bubble. Plot the valuation of alternative energy firms in the 2000s alongside the NASDAQ's rise and fall a decade before, and you see the same shape:

The 1990s had one big idea: *the internet is going to be big*. But too many internet companies had exactly that same idea and no others. An entrepreneur can't benefit from macroscale insight unless his own plans begin at the micro-scale. Cleantech companies faced the same problem: no matter how much the world needs energy, only a firm that offers a superior solution for a specific energy problem can make money. No sector will ever be so important that merely participating in it will be enough to build a great company.

The tech bubble was far bigger than cleantech and the crash even more painful. But the dream of the '90s turned out to be right: skeptics who doubted that the internet would fundamentally change publishing or retail sales or everyday social life looked prescient in 2001, but they seem comically foolish today. Could successful energy startups be founded after the cleantech crash just as Web 2.0 startups successfully launched amid the debris of the dot-coms? The macro need for energy solutions is still real. But a valuable business must start by finding a niche and dominating a small market. Facebook started as a service for just one university campus before it spread to other schools and then the entire world. Finding small markets for energy solutions will be tricky—you could aim to replace diesel as a power source for remote islands, or maybe build modular reactors for quick deployment at military installations in hostile territories. Paradoxically, the challenge for the entrepreneurs who will create Energy 2.0 is to think small.

# 14

## THE FOUNDER'S PARADOX

O F THE SIX PEOPLE who started PayPal, four had built bombs in high school.

Five were just 23 years old—or younger. Four of us had been born outside the United States. Three had escaped here from communist countries: Yu Pan from China, Luke Nosek from Poland, and Max Levchin from Soviet Ukraine. Building bombs was not what kids normally did in those countries at that time.

The six of us could have been seen as eccentric. My first-ever conversation with Luke was about how he'd just signed up for cryonics, to be frozen upon death in hope of medical resurrection. Max claimed to be without a country and proud of it: his family was put into diplomatic limbo when the USSR collapsed while they were escaping to the U.S. Russ Simmons had escaped from a trailer park to the top math and science magnet school in Illinois. Only Ken Howery fit the stereotype of a privileged American childhood: he

The PayPal Team in 1999

was PayPal's sole Eagle Scout. But Kenny's peers thought he was crazy to join the rest of us and make just one-third of the salary he had been offered by a big bank. So even he wasn't entirely normal.

Are all founders unusual people? Or do we just tend to remember and exaggerate whatever is most unusual about them? More important, which personal traits actually matter in a founder? This chapter is about why it's more powerful but at the same time more dangerous for a company to be led by a distinctive individual instead of an interchangeable manager.

## THE DIFFERENCE ENGINE

Some people are strong, some are weak, some are geniuses, some are dullards—but most people are in the middle. Plot where everyone falls and you'll see a bell curve:

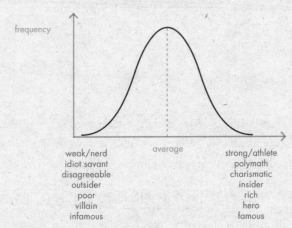

## NORMAL DISTRIBUTION OF TRAITS

frequency

average

weak/nerd
idiot savant
disagreeable
outsider
poor
villain
infamous

strong/athlete
polymath
charismatic
insider
rich
hero
famous

Since so many founders seem to have extreme traits, you might guess that a plot showing only founders' traits would have fatter tails with more people at either end.

But that doesn't capture the strangest thing about founders. Normally we expect opposite traits to be mutually exclusive: a normal person can't be both rich and poor at the same time, for instance. But it happens all the time to founders: startup CEOs can be cash poor but millionaires on paper. They may oscillate between sullen jerkiness and appealing charisma. Almost all successful entrepreneurs are simultaneously insid-

## FAT-TAILED DISTRIBUTION

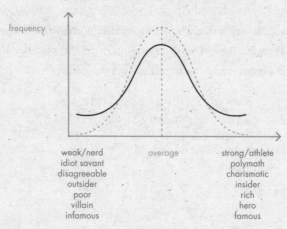

ers and outsiders. And when they do succeed, they attract both fame and infamy. When you plot them out, founders' traits appear to follow an inverse normal distribution:

## THE FOUNDER DISTRIBUTION

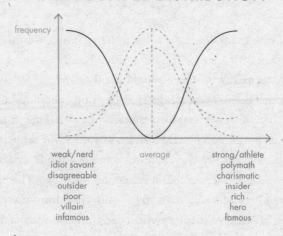

Where does this strange and extreme combination of traits come from? They could be present from birth (nature) or acquired from an individual's environment (nurture). But perhaps founders aren't really as extreme as they appear. Might they strategically exaggerate certain qualities? Or is it possible that everyone else exaggerates them? All of these effects can be present at the same time, and whenever present they powerfully reinforce each other. The cycle usually starts with unusual people and ends with them acting and seeming even more unusual:

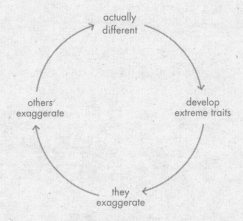

As an example, take Sir Richard Branson, the billionaire founder of the Virgin Group. He could be described as a natural entrepreneur: Branson started his first business at age 16, and at just 22 he founded Virgin Records. But other aspects of his renown—the trademark lion's mane hairstyle, for example—are less natural: one suspects he wasn't born with that exact look. As Branson has cultivated his other extreme traits (Is kiteboarding with naked supermodels a PR stunt? Just a guy having fun? Both?), the media has eagerly enthroned him: Branson is "The Virgin King," "The Undisputed King of PR," "The King of Branding," and "The King of the Desert and Space." When Virgin Atlantic Airways began serving passengers drinks with ice cubes shaped like Branson's face, he became "The Ice King."

Is Branson just a normal businessman who happens to be lionized by the media with the help of a good PR team? Or is he himself a born branding genius rightly singled out by the journalists he is so good at manipulating? It's hard to tell—maybe he's both.

Another example is Sean Parker, who started out with the ultimate outsider status: criminal. Sean was a careful hacker in high school. But his father decided that Sean was spending too much time on the computer for a 16-year-old, so one day he took away Sean's keyboard mid-hack. Sean couldn't log out; the FBI noticed; soon federal agents were placing him under arrest.

Sean got off easy since he was a minor; if anything, the episode emboldened him. Three years later, he co-founded Napster. The peer-to-peer file sharing service amassed 10 million users in its first year, making it one of the fastest-growing businesses of all time. But the record companies sued and a federal judge ordered it shut down 20 months after opening. After a whirlwind period at the center, Sean was back to being an outsider again.

Then came Facebook. Sean met Mark Zuckerberg in 2004, helped negotiate Facebook's first funding, and became the company's founding president. He had to step down in 2005 amid allegations of drug use, but this only enhanced his notoriety. Ever since Justin Timberlake portrayed him in *The Social Network,* Sean has been perceived as one of the coolest people in America. JT is still more famous, but when he visits Silicon Valley, people ask if he's Sean Parker.

The most famous people in the world are founders, too: instead of a company, every celebrity founds and cultivates a personal brand. Lady Gaga, for example, became one of the most influential living people. But is she even a real person? Her real name isn't a secret, but almost no one knows or cares what it is. She wears costumes so bizarre as to put any other wearer at risk of an involuntary psychiatric hold. Gaga would have you believe that she was "born this way"— the title of both her second album and its lead track. But no one is born looking like a zombie with horns coming out of her head: Gaga must therefore be a self-manufactured myth. Then again, what kind of person would do this to herself? Certainly nobody normal. So perhaps Gaga really *was* born that way.

## WHERE KINGS COME FROM

Extreme founder figures are not new in human affairs. Classical mythology is full of them. Oedipus is the paradigmatic insider/outsider: he was abandoned as an infant and ended up in a foreign land, but he was a brilliant king and smart enough to solve the riddle of the Sphinx.

Romulus and Remus were born of royal blood and abandoned as orphans. When they discovered their pedigree, they decided to found a city. But they couldn't agree on where to put it. When Remus crossed the boundary that Romulus had decided was the edge of Rome, Romulus killed him, declaring: "So perish every one that shall hereafter leap over my wall." Law-maker *and* law-breaker, criminal outlaw *and* king who defined Rome, Romulus was a self-contradictory insider/outsider.

Normal people aren't like Oedipus or Romulus. Whatever those individuals were actually like in life, the mythologized versions of them remember only the extremes. But why was it so important for archaic cultures to remember extraordinary people?

The famous and infamous have always served as vessels for public sentiment: they're praised amid prosperity and blamed for misfortune. Primitive societies faced one fundamental problem above all: they would be torn apart by conflict if they didn't have a way to stop it. So whenever plagues, disasters, or violent rivalries threatened the peace, it was beneficial for the society to place the entire blame on a single person, someone everybody could agree on: a scapegoat.

Who makes an effective scapegoat? Like founders, scapegoats are extreme and contradictory figures. On the one hand, a scapegoat is necessarily weak; he is powerless to stop his own victimization. On the other hand, as the one who can defuse conflict by taking the blame, he is the most powerful member of the community.

Before execution, scapegoats were often worshipped like deities. The Aztecs considered their victims to be earthly forms of the gods to whom they were sacrificed. You would be dressed in fine clothes and feast royally until your brief reign ended and they cut your heart out. These are the roots of monarchy: every king was a living god, and every god a murdered king. Perhaps every modern king is just a scapegoat who has managed to delay his own execution.

## AMERICAN ROYALTY

Celebrities are supposedly "American royalty." We even grant titles to our favorite performers: Elvis Presley was the king of rock. Michael Jackson was the king of pop. Britney Spears was the pop princess.

Until they weren't. Elvis self-destructed in the '70s and died alone, overweight, sitting on his toilet. Today, his impersonators are fat and sketchy, not lean and cool. Michael Jackson went from beloved child star to an erratic, physically repulsive, drug-addicted shell of his former self; the world reveled in the details of his trials. Britney's story is the most dramatic of all. We created her from nothing, elevating her to superstardom as a teenager. But then everything fell off the tracks: witness the shaved head, the over- and under-eating scandals, and the highly publicized court case to take away her children. Was she always a little bit crazy? Did the publicity just get to her? Or did she do it all to get more?

For some fallen stars, death brings resurrection. So many popular musicians have died at age 27—Janis Joplin, Jimi Hendrix, Jim Morrison, and Kurt Cobain, for example—that this set has become immortalized as the "27 Club." Before she joined the club in 2011, Amy Winehouse sang: "They tried to make me go to rehab, but I said, 'No, no, no.'" Maybe rehab seemed so unattractive because it blocked the path to immortality. Perhaps the only way to be a rock god forever is to die an early death.

We alternately worship and despise technology founders just as we do celebrities. Howard Hughes's arc from fame to pity is the most dramatic of any 20th-century tech founder. He was born wealthy, but he was always more interested in engineering than luxury. He built Houston's first radio transmitter at the age of 11. The year after that he built the city's first motorcycle. By age 30 he'd made nine commercially successful movies at a time when Hollywood was on the technological frontier. But Hughes was even more famous for his parallel career in aviation. He designed planes, produced them, and piloted them himself. Hughes set world records for top airspeed, fastest transcontinental flight, and fastest flight around the world.

Hughes was obsessed with flying higher than everyone else. He liked to remind people that he was a mere mortal, not a Greek god—something that mortals say only when they want to invite comparisons to gods. Hughes was "a man to whom you cannot apply the same standards as you can to you and me," his lawyer once argued in federal court. Hughes paid the lawyer to say that, but according to the *New York Times* there was "no dispute on this point from judge or jury." When Hughes was awarded the Congressional Gold Medal in 1939 for his achievements in aviation, he didn't even show up to claim it—years later President Truman found it in the White House and mailed it to him.

The beginning of Hughes's end came in 1946, when he suffered his third and worst plane crash. Had he died then, he would have been remembered forever as one of the most dashing and successful Americans of all time. But he survived—barely. He became obsessive-compulsive, addicted to painkillers, and withdrew from the public to spend the last 30 years of his life in self-imposed solitary confinement. Hughes had always acted a little crazy, on the theory that fewer people would want to bother a crazy person. But when his crazy act turned into a crazy life, he became an object of pity as much as awe.

More recently, Bill Gates has shown how highly visible success can attract highly focused attacks. Gates embodied the founder archetype: he was simultaneously an awkward and nerdy college-dropout outsider and the world's wealthiest insider. Did he choose his geeky eyeglasses strategically, to build up a distinctive persona? Or, in his incurable nerdiness, did his geek glasses choose him? It's hard to know. But his dominance was undeniable: Microsoft's Windows claimed a

90% share of the market for operating systems in 2000. That year Peter Jennings could plausibly ask, "Who is more important in the world today: Bill Clinton or Bill Gates? I don't know. It's a good question."

The U.S. Department of Justice didn't limit itself to rhetorical questions; they opened an investigation and sued Microsoft for "anticompetitive conduct." In June 2000 a court ordered that Microsoft be broken apart. Gates had stepped down as CEO of Microsoft six months earlier, having been forced to spend most of his time responding to legal threats instead of building new technology. A court of appeals later overturned the breakup order, and Microsoft reached a settlement with the government in 2001. But by then Gates's enemies had already deprived his company of the full engagement of its founder, and Microsoft entered an era of relative stagnation. Today Gates is better known as a philanthropist than a technologist.

## THE RETURN OF THE KING

Just as the legal attack on Microsoft was ending Bill Gates's dominance, Steve Jobs's return to Apple demonstrated the irreplaceable value of a company's founder. In some ways, Steve Jobs and Bill Gates were opposites. Jobs was an artist, preferred closed systems, and spent his time thinking about great products above all else; Gates was a businessman, kept his products open, and wanted to run the world. But both were insider/outsiders, and both pushed the companies they started to achievements that nobody else would have been able to match.

A college dropout who walked around barefoot and refused to shower, Jobs was also the insider of his own personality cult. He could act charismatic or crazy, perhaps according to his mood or perhaps according to his calculations; it's hard to believe that such weird practices as apple-only diets weren't part of a larger strategy. But all this eccentricity backfired on him in 1985: Apple's board effectively kicked Jobs out of his own company when he clashed with the professional CEO brought in to provide adult supervision.

Jobs's return to Apple 12 years later shows how the most important task in business—the creation of new value—cannot be reduced to a formula and applied by professionals. When he was hired as interim CEO of Apple in 1997, the impeccably credentialed executives who preceded him had steered the company nearly to bankruptcy. That year Michael Dell famously said of Apple, "What would I do? I'd shut it down and give the money back to the shareholders." Instead Jobs introduced the iPod (2001), the iPhone (2007), and the iPad (2010) before he had to resign in 2011 because of poor health. By the following year Apple was the single most valuable company in the world.

Apple's value crucially depended on the singular vision of a particular person. This hints at the strange way in which the companies that create new technology often resemble feudal monarchies rather than organizations that are supposedly more "modern." A unique founder can make authoritative decisions, inspire strong personal loyalty, and plan ahead for decades. Paradoxically, impersonal bureaucracies staffed by trained professionals can last longer than any lifetime, but they usually act with short time horizons.

The lesson for business is that we need founders. If anything, we should be more tolerant of founders who seem strange or extreme; we need unusual individuals to lead companies beyond mere incrementalism.

The lesson for founders is that individual prominence and

adulation can never be enjoyed except on the condition that it may be exchanged for individual notoriety and demonization at any moment—so be careful.

Above all, don't overestimate your own power as an individual. Founders are important not because they are the only ones whose work has value, but rather because a great founder can bring out the best work from everybody at his company. That we need individual founders in all their peculiarity does not mean that we are called to worship Ayn Randian "prime movers" who claim to be independent of everybody around them. In this respect Rand was a merely half-great writer: her villains were real, but her heroes were fake. There is no Galt's Gulch. There is no secession from society. To believe yourself invested with divine self-sufficiency is not the mark of a strong individual, but of a person who has mistaken the crowd's worship—or jeering—for the truth. The single greatest danger for a founder is to become so certain of his own myth that he loses his mind. But an equally insidious danger for every business is to lose all sense of myth and mistake disenchantment for wisdom.

*Conclusion*

# STAGNATION OR SINGULARITY?

IF EVEN THE MOST FARSIGHTED founders cannot plan beyond the next 20 to 30 years, is there anything to say about the very distant future? We don't know anything specific, but we can make out the broad contours. Philosopher Nick Bostrom describes four possible patterns for the future of humanity.

The ancients saw all of history as a neverending alternation between prosperity and ruin. Only recently have people dared to hope that we might permanently escape misfortune, and it's still possible to wonder whether the stability we take for granted will last.

## RECURRENT COLLAPSE

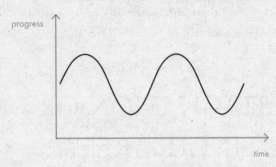

However, we usually suppress our doubts. Conventional wisdom seems to assume instead that the whole world will converge toward a plateau of development similar to the life of the richest countries today. In this scenario, the future will look a lot like the present.

## PLATEAU

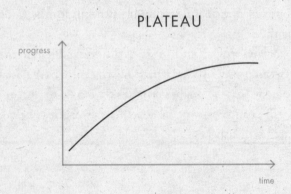

Given the interconnected geography of the contemporary world and the unprecedented destructive power of modern weaponry, it's hard not to ask whether a large-scale social disaster could be contained were it to occur. This is what fuels our fears of the third possible scenario: a collapse so devastating that we won't survive it.

## EXTINCTION

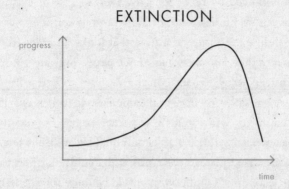

The last of the four possibilities is the hardest one to imagine: accelerating takeoff toward a much better future. The end result of such a breakthrough could take a number of forms, but any one of them would be so different from the present as to defy description.

## TAKEOFF

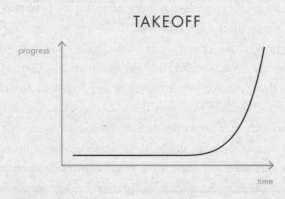

Which of the four will it be?

Recurrent collapse seems unlikely: the knowledge underlying civilization is so widespread today that complete annihilation would be more probable than a long period of darkness followed by recovery. However, in case of extinction, there is no human future of any kind to consider.

If we define the future as a time that looks different from the present, then most people aren't expecting any future at all; instead, they expect coming decades to bring more globalization, convergence, and sameness. In this scenario, poorer countries will catch up to richer countries, and the world as a whole will reach an economic plateau. But even if a truly globalized plateau were possible, could it last? In the best case, economic competition would be more intense than ever before for every single person and firm on the planet.

However, when you add competition to consume scarce resources, it's hard to see how a global plateau could last indefinitely. Without new technology to relieve competitive pressures, stagnation is likely to erupt into conflict. In case of conflict on a global scale, stagnation collapses into extinction.

That leaves the fourth scenario, in which we create new technology to make a much better future. The most dramatic version of this outcome is called the Singularity, an attempt to name the imagined result of new technologies so powerful as to transcend the current limits of our understanding. Ray Kurzweil, the best-known Singularitarian, starts from Moore's law and traces exponential growth trends in dozens of fields, confidently projecting a future of superhuman artificial intelligence. According to Kurzweil, "the Singularity is near," it's inevitable, and all we have to do is prepare ourselves to accept it.

But no matter how many trends can be traced, the future won't happen on its own. What the Singularity would look like matters less than the stark choice we face today between the two most likely scenarios: nothing or something. It's up to us. We cannot take for granted that the future will be better, and that means we need to work to create it today.

Whether we achieve the Singularity on a cosmic scale is perhaps less important than whether we seize the unique opportunities we have to do new things in our own working lives. Everything important to us—the universe, the planet, the country, your company, your life, and this very moment—is singular.

Our task today is to find singular ways to create the new things that will make the future not just different, but better—to go from 0 to 1. The essential first step is to think for yourself. Only by seeing our world anew, as fresh and strange as it was to the ancients who saw it first, can we both re-create it and preserve it for the future.

# Acknowledgments

Jimmy Kaltreider for helping to think this book through.

Rob Morrow, Scott Nolan, and Michael Solana for co-creating the Stanford class from which we started.

Chris Parris-Lamb, Tina Constable, David Drake, Talia Krohn, and Jeremiah Hall for skillfully guiding us to publication.

Everyone at Thiel Capital, Founders Fund, Mithril, and the Thiel Foundation for working hard and smart.

Onward.

# Illustration Credits

The illustrations in this book were drawn by Matt Buck, based on the following images:

page 96:    *Unabomber,* Jeanne Boylan/FBI composite sketch
page 96:    *Hipster,* Matt Buck
page 161:   *Brian Harrison,* Business Wire
page 161:   *Elon Musk,* Sebastian Blanco and Bloomberg/Getty Images
page 178:   *Richard Branson,* Ethan Miller/Getty Images
page 179:   *Sean Parker,* Aaron Fulkerson, flickr user Roebot, used under CC BY-SA
page 182:   *Elvis Presley,* Michael Ochs Archives/Getty Images
page 182:   *Michael Jackson,* Frank Edwards/Getty Images
page 182:   *Britney Spears,* Kevin Mazur Archive 1/WireImage
page 183:   *Elvis Presley,* Tom Wargacki/WireImage
page 183:   *Michael Jackson,* David LEFRANC/Gamma-Rapho via Getty Images
page 183:   *Britney Spears,* New York Daily News Archive via Getty Images
page 183:   *Janis Joplin,* Albert B. Grossman and David Gahr/Getty Images
page 183:   *Jim Morrison,* Elektra Records and CBS via Getty Images
page 183:   *Kurt Cobain,* Frank Micelotta/Stringer/Getty Images
page 183:   *Amy Winehouse,* flickr user teakwood, used under CC BY-SA
page 185:   *Howard Hughes,* Bettmann/CORBIS
page 185:   *magazine cover, TIME,* a division of Time Inc.
page 186:   *Bill Gates,* Doug Wilson/CORBIS
page 186:   *magazine cover, Newsweek*
page 187:   *Steve Jobs, 1984,* Norman Seeff
page 187:   *Steve Jobs, 2004,* Contour by Getty Images

# Index

Page numbers in *italics* refer to illustrations.

# About the Authors

**Peter Thiel** is an entrepreneur and investor. He started PayPal in 1998, led it as CEO, and took it public in 2002, defining a new era of fast and secure online commerce. In 2004 he made the first outside investment in Facebook, where he serves as a director. The same year he launched Palantir Technologies, a software company that harnesses computers to empower human analysts in fields like national security and global finance. He has provided early funding for LinkedIn, Yelp, and dozens of successful technology startups, many run by former colleagues who have been dubbed the "PayPal Mafia." He is a partner at Founders Fund, a Silicon Valley venture capital firm that has funded companies like SpaceX and Airbnb. He started the Thiel Fellowship, which ignited a national debate by encouraging young people to put learning before schooling, and he leads the Thiel Foundation, which works to advance technological progress and long-term thinking about the future.

**Blake Masters** was a student at Stanford Law School in 2012 when his detailed notes on Peter's class "Computer Science 183: Startup" became an internet sensation. He went on to co-found Judicata, a legal research technology startup.